Corporate Research Laboratories and the History of Innovation

At the beginning of the twentieth century, American corporations in the chemical and electrical industries began establishing industrial research laboratories. Some went on to become world-famous not only for their scientific and technological breakthroughs but also for the new union of science and industry they represented. Innovative ideas do not simply appear out of the blue and spread on their own merit. Rather, the laboratory's diffusion takes place in a cultural context that goes beyond corporate capital and technological change.

Using discourse analysis as a method to comprehensively capture the organizational field of the early American R&D laboratories from 1870 to 1930, this book uncovers the collective meanings associated with the industrial laboratory. Meanings such as what and where a laboratory is supposed to be, who the scientist is, and what it means to practice science provided cultural resources that made the transfer of the laboratory from academic science into an industrial setting possible by rendering such meanings understandable and operable to big business and organizational entrepreneurs fighting for hegemony in a rapidly evolving market. It analyzes not only the corporations that established laboratories in the United States but also their contexts – economic, political, and especially scientific – showing how "the industrial laboratory" was transformed from an organizational novelty into an expected institution in less than two decades.

This book will be of interest to researchers, academics, historians, and students in the fields of organizational change, discourse studies, the management of technology and innovation, as well as business and management history.

David M. Pithan is a Programme Officer for Strategy and Policy at the German Research Foundation.

Routledge Studies in Management, Organizations and Society

This series presents innovative work grounded in new realities, addressing issues crucial to an understanding of the contemporary world. This is the world of organised societies, where boundaries between formal and informal, public and private, local and global organizations have been displaced or have vanished, along with other nineteenth-century dichotomies and oppositions. Management, apart from becoming a specialized profession for a growing number of people, is an everyday activity for most members of modern societies.

Similarly, at the level of enquiry, culture and technology, and literature and economics, can no longer be conceived as isolated intellectual fields; conventional canons and established mainstreams are contested. *Management, Organizations and Society* addresses these contemporary dynamics of transformation in a manner that transcends disciplinary boundaries, with books that will appeal to researchers, students, and practitioners alike.

Recent titles in this series include:

Founders and Organizational Development
The Etiology and Theory of Founder's Syndrome
Stephen R. Block and Katrina Miller-Stevens

Profitability, Productivity, and Sustainability
Organizational Behavior and Strategic Alignment
Dennis N. Onyama

Corporate Research Laboratories and the History of Innovation
David M. Pithan

Change and Development in Organisations
Towards Consciousness, Humanity and Innovation
Edited by Ricardo Chiva

Towards the Compassionate University
From Golden Thread to Global Impact
Edited by Kathryn Waddington

Corporate Research Laboratories and the History of Innovation

David M. Pithan

Routledge
Taylor & Francis Group
NEW YORK AND LONDON

First published 2022
by Routledge
605 Third Avenue, New York, NY 10158

and by Routledge
2 Park Square, Milton Park, Abingdon, Oxon, OX14 4RN

Routledge is an imprint of the Taylor & Francis Group, an informa business

Library of Congress Cataloging-in-Publication Data
A catalog record has been requested for this book

ISBN: 978-0-367-47660-1 (hbk)
ISBN: 978-0-367-53840-8 (pbk)
ISBN: 978-1-003-08347-4 (ebk)

Typeset in Sabon
by MPS Limited, Dehradun

Contents

List of Illustrations

Figures

Tables

Acknowledgments

This book is a revised and streamlined version of my PhD thesis. First and foremost, I would like to thank my PhD advisor, Thomas Heinze, who supported me throughout all the stages of this project. Further thanks for all the support I received from the team of Organizational Sociology at Wuppertal University, namely Steffi Heinecke, Sabrina Petersohn, Olof Hallonsten, Arlette Jappe, Dennis Köthemann, and Joschka Baum.

Thanks to the kind and helpful staff at the Science History Institute (Philadelphia, PA), at the Hagley Museum and Library (Wilmington, DE), as well as at the Kodak archives at the University of Rochester and the George Eastman Museum (Rochester, NY), for opening the gates of their archives for me and my project.

I am very grateful for countless critical and inspirational discussions with my colleagues at the Institute of Sociology and the Interdisciplinary Centre for Science and Technology (IZWT), both at Wuppertal University, especially Elena Kunadt, Julia Heuer, Aytüre Türkyilmaz, Miriam Böttner, and Roman Pauli.

Lastly, thanks to my family and friends for supporting me throughout the way, and a very special thanks to those who proofread, sat with me through long brainstorming sessions, and helped me figure things out: Michael Rose, Vincent Gengnagel, Heiner Heiland, and Andreas Kallert.

List of Abbreviations

AAAS	American Association for the Advancement of Science
ABT	American Bell Telephone Company
ACS	American Chemical Society
AES	American Electrochemical Society
AIC	American Institute of Chemists
AIChE	American Institute of Chemical Engineers
AJS	American Journal of Science
AT&T	American Telephone & Telegraph Company
CDA	Critical Discourse Analysis
DIR	Directors of Industrial Research
EGE	Edison General Electric
GE	General Electric
IUPAC	International Union of Pure and Applied Chemistry
JACS	Journal of the American Chemical Society
JIEC	Journal of Industrial & Engineering Chemistry
MCA	Manufacturing Chemists' Association
NAM	National Association of Manufacturers
NAS	National Academy of Sciences
NCB	Naval Consulting Board
NRC	National Research Council
NRE	National Research Endowment
ODA	Organizational Discourse Analysis
R&D	Research and Development
SoCI	Society for Chemical Industry
STS	Science and Technology Studies
TH	Thomson-Houston
USDA	United States Department of Agriculture
WU	Western Union
WWI	World War I

1 Introduction

This is a book about the history of ideas, and about one idea in particular: setting up a laboratory within the confines of a business enterprise, bringing the material locus of science into a world that seems to be pursuing entirely different goals, subject to an alien logic. When untangling the complex web of traces that ideas leave when they travel, one will quickly notice that the history of the industrial laboratory is also a history of science, and history not only of its contents, practitioners, and institutions but also of the many meanings associated with the concept of science over the centuries. These meanings transformed what practicing science – and practicing science in industry – signified at different times, and in different contexts. While the industrial laboratory may be taken for granted today, conjuring images of eponymous triumphs, chemists clad in white coats, or where superheroes and supervillains are made in our fiction, understanding it as a specific materialization of the practice of science clearly highlights its contingent nature. But this is not a book about laboratory lives and the very nature of the scientific enterprise. Instead, it is about how the laboratory came to be so pervasive and taken for granted. In this way, the third kind of history is revealed, that of organizations. Modern society is increasingly characterized as pervaded by organizations, so it is only fitting to scrutinize one of those organizational entities often seen as a prime example of modern corporations and their complex interrelations with each other – and their environments – from a sociological perspective on organizations.

Of course, numerous scholars have already covered the histories of the early R&D laboratories in various levels of detail. Several waves of scholarly interest have swept over corporate archives, ranging from early – rather celebratory – histories, to the organizational histories of the 1980s, focusing on the most prolific and illustrious laboratories. Each wave reframed these labs, uncovering new aspects of laboratories in industry, the inner workings of science, and big corporations. This is why so much is known about some of the early labs, such as their specific triumphs and failures, and their structural shifts and transformations. But, as Lucier notes, interest in corporate labs has recently ceased: "Perhaps it was a sense that there was nothing left of theoretical interest to say about industrial research ..., which along with other institutional histories have looked rather dull and outdated since the cultural turn in history" (Lucier 2016, 25). So, since so much appears to be said already about this topic, what exactly warrants *another* look at this kind of laboratory?

The perspective this book takes is effectively informed by an interest in the diffusion of innovations, combined with a sociological hunch: that in order to spread successfully, new ideas need to be legitimated in some way. Legitimacy, here, means appearing congruent – fitting and natural – with the existing order. Of course, ideas do not travel on their own, and similarly, they need actors championing their acceptance. Hence, diffusing ideas are subject to a multitude of transformations and translations that fundamentally change both their form and content in order to fit new environments and organizational contexts. It is precisely this intersection within which the main analytical thrust of this study lies: how new ideas are made to fit – through individual acts of translation – to greater collective layers of meaning, thereby providing legitimacy, which I call discourse. These discourses construct social reality in the first place: its categories, actors, and possibilities for the agency. Thus, to understand how scientific laboratories found a place in industrial corporations, it first needs to be known what the laboratory, science, and the scientist meant to organizational actors of the time. How these concepts were translated – discursively re-constructed – in order to fit different local contexts and grander discursive shifts and transformations is at the heart of this successful process of diffusion.

In this unique convergence of the history of ideas, science, and the sociology of organizations, the constructive effects of language on organizational reality are key to the way that this new idea was naturalized and made salient. Foregoing grand explanations that see the industrial laboratory as an inevitable organizational outcome of modernization and the general rationalization of economic life, the goal here is to combine detailed history with contextual factors in order to paint a more complete picture: "Because laboratories are so integrally a part of their times and places, lab history is of necessity also social history" (Kohler 2008, 765). A theoretical framework combining organizational theory with discursive methods is used, working toward a cultural explanation of how laboratories in industry came to be institutionalized. Hence, going beyond existing studies on corporate R&D that center around organizational, technical, or legal aspects, my study will include the ideational environment of chemical corporations and how its characteristics conditioned organizational agency. In order to gain access to historical scholarship on the topic, as well as texts that were later used in the discourse analysis, I spent four months at the Chemical Heritage Foundation in Philadelphia, PA, in the fall of 2016. During this time, I also visited the Du Pont archives at the Hagley Museum and Library in

Wilmington, DE, as well as the Kodak archives at the University of Rochester and the George Eastman Museum in Rochester, NY. These research visits proved instrumental in filling some holes in the histories of the "research pioneers", as well as adequately mapping the field's discourses.

This book will focus on industrial research laboratories in the United States only, and specifically in the US chemical industries. In the beginning, the hypothesis arose that the adoption of R&D in America was the result of the highly successful German template of laboratories in the chemical industries. But soon after it was revealed that American firms followed their own paths in this venture, nullifying explanations based on visibility and mimesis. A comparative study surely would have been interesting, but the exploratory approach and hybrid method employed here set limitations. Mapping two organizational fields' discourses with wildly different histories would have been excessive and may have diluted the analytical clarity of either analysis. Consequently, the focus was shifted toward the unique economical, political, and cultural situation in the United States.

Anyone familiar with previous scholarship on industrial research laboratories and the history of American science will know how their stories are often informed by perspectives interested in the types of science done within the laboratories. These types may be called pure or applied science, or basic and applied research, among many other descriptors. Discussions found within such approaches are often concerned with the relationships between these types, or how pure science could be practiced in impure places, such as anywhere outside of academia. Using an analytical perspective here that steers clear of any such preconceived notions was a conscious, purposeful decision. Any assumptions regarding the types of science, the motives of its practitioners, and their "proper" locations presuppose science in fixed, distorted ways that would run counter to a framework and epistemology that puts the constructive effects of language front and center. In addition, focusing on an ontology of science and its related ideologies would run the risk of obscuring what actually happened, as Gieryn put it very succinctly: "The sociological question is not whether science is really pure or impure or both, but rather how its borders and territories are flexibly and discursively mapped out in pursuit of some observed or inferred ambition – and with what consequences, and for whom?" (Gieryn 1999, 23). Hence, when mapping out the pathways, the laboratory took from organizational novelty to institution, no presumptions about any types of science involved shall be made; instead, the actors of the time who were involved in its spread shall be heard.

Lastly, a note on the designations used in this book. In the following, I will variously speak of laboratories, industrial research laboratories, R&D labs, and other labels. As the history of organized research in industry in America laid out below will illustrate, a plethora of terms existed for the organizational unit, and categories such as "research" and "development" were in no ways fixed (or even in existence) at the beginning of the twentieth century. Different names for these laboratories will thus only be used for stylistic reasons unless otherwise noted, and in no way to make assumptions about what kind of work was done there. In the end, what this book is interested in is how modern corporations came to have a distinct unit called "research laboratory", no matter its place in the organizational structure or what specifically happened within its laboratory walls.

2 Theory

Diffusion of Innovations

In the following, I develop a theoretical framework for analyzing the emergence and spread of the industrial research laboratory, which brings together various perspectives on innovations and organizations from different strands of sociology. These perspectives serve not only to put the diffusion of ideas into a conceptual model but also to provide a methodical starting point for the discourse analysis. Of course, many other framings for understanding R&D laboratories could have been chosen, but combining diffusion with organizational theory adds a new angle to both the story of the industrial laboratory and to the diffusion of innovations in organizational fields by focusing on the role that an innovation's fit within the surrounding environment plays. To understand how new ideas, practices, or things spread successfully, one needs to account for cultural variables that serve as anchoring points for traveling ideas. Building a theoretical framework for such an analysis will be undertaken in three steps: First, I examine historical and conceptual developments in the literature of diffusion of innovations, with a particular focus on cultural variables in social systems in order to better understand either how innovations fit or are made to fit. Second, since the history of the R&D laboratory is about organizations, I use elements from organizational institutionalism to expand the toolkit of diffusion studies, especially by using the organizational field concept that improves our understanding of what conditions organizational agency within organizational fields. Third, the concept of discourse – and the method of discourse analysis – will be introduced as a way to analyze the layers of meaning in a given field. After this theoretical triple jump, the research program that follows the theoretical framework is outlined.

What a Diffusion Perspective Can Add

What do the theoretical perspective and methodical toolkit of diffusion studies add to the understanding of the emergence of the industrial research laboratory – and, vice versa, how can this example contribute to the already extensive body of literature on diffusion? As it turns out, the curious case of the birth of the organizational entity that is known today as the Research and Development Laboratory (or R&D Lab, for short) has already attracted considerable scholarly attention, having been addressed from a variety of angles. Historians of science and technology (Reich 1980; Wise 1985; Hounshell & Smith 1988) explored the singular histories of the pioneer laboratories, their eponymous directors, their scientific triumphs, and the external and internal resistances – ranging from a rapidly changing legal and economic environment to the challenges of doing "scientific research" in an industrial setting – that needed to be overcome in order to shape a space for science in the modern corporation. In contrast, economic sociologists (Chandler 1977, 1990b; Fligstein 1990) put the R&D Laboratory in the context of the larger shifts of corporations' structures and managerial logic happening at the time, linking environmental pressures to the internal demands of the giant corporations that ultimately brought about an organizational structure that included a laboratory – structure following strategy, an expression of managerial ideologies. Furthermore, innovation economists (Mowery 1981; Lamoreaux & Sokoloff 1999) added to our understanding by including quantitative indicators in a framework of transactional costs and (technological) markets, underlining the roles played by factors such as sector, firm size, and capital in the establishment of laboratories. Theoretical perspectives on corporate research laboratories are disassembled and their various explanatory parts scrutinized, in a history of the early US R&D laboratories sketched later (Chapter 3).

Clearly, previous scholarship can answer the question "What happened?": The Industrial Research Laboratory was created within large American corporations around 1900, primarily in the chemical and electrical industries. The laboratory as an organizational entity then spread through these and adjacent industries, such as the pharmaceutical industry (Liebenau 1985; Furman & MacGarvie 2007). This process can certainly be regarded as an account of organizational change, yet change that was not triggered by new legislation demanding the establishment of laboratory facilities, but rather by a unique combination of factors making the spread of this organizational entity through a population of organizations possible. As such, the emergence and spread of the industrial research laboratory can be framed as a process of diffusion – diffusion being "the most general and abstract term we have for this sort of process, embracing contagion, mimicry, social learning, organized dissemination, and other family members" (Strang & Soule 1998, 266). In particular, this framing adds the analytical categories developed over decades of diffusion research to describe the emergence and spread of R&D laboratories not as heroic entrepreneurs enacting change from within organizations, as the historic accounts would suggest, or as field-level conditions forcing an inevitable outcome. Instead, the toolbox to be utilized here (cf. Rogers 2003, also Mahajan & Peterson 1985; Palloni 2001; Wejnert 2002) offers ways of accounting for individual action within a larger environment that, when viewed a posteriori, takes the shape of a cascading mechanism, of something spreading throughout a social system.

Traditionally, diffusion studies include four elements in their analytical framework: an *innovation* that spreads via *channels of communication*, over *time*, throughout a *social system* (Rogers 2003, 1ff). The element of time is defined by the empirical reality of the spread of R&D, which happened in the first three decades of the twentieth century. The specific channels of communication themselves are of lesser interest to this study: First, they are subordinated to a social structure that will be scrutinized in detail – patterns of communication now a by-product of patterns of interaction. Second, when analyzing texts and their respective speakers methodically, channels of communication are included in the three-dimensional approach to discourse utilized here. Some definitional trouble starts when accounting for the specific item that diffuses. In Roger's authoritative review of diffusion research (2003) that item is termed an innovation – indeed a term usually occurring in conjunction with diffusion: "The home territory of diffusion is the innovation" (Strang & Soule 1998, 267). Innovation is defined as "an idea, practice, or object that is perceived as new by an individual or other unit of adoption" (Rogers 2003, 12). Particular emphasis is laid on the perception of novelty, regardless of "objective" newness. The notion of innovation encompasses a variety of items, such as ideas or practices as noted previously, also material things such as technologies (for an introduction, see Fagerberg 2005).

Of course, specifically when dealing with a historically diffuse concept such as "innovation", a word of caution is in order. The corporate R&D laboratory can be deemed an innovation only with the benefit of hindsight – by virtue of analyzing its spread more than a hundred years later. Today, it is known that the laboratory became an institutionalized part of the modern corporation. Yet terming the R&D lab an innovation from the outset may occlude the reasons for its successful diffusion on a conceptual level by assuming inherent qualities that assured its diffusion. Furthermore, one runs the risk of falling prey to a pro-innovation bias that has plagued the scholarship of innovation: since only those ideas, practices, or things that "make it" can be analyzed, any hypothetical sample of things that diffuse is heavily skewed toward successful innovations. Accounts of failed diffusion – and failed innovations – are few and far between (cf. Rogers 2003, 1–5, 8–11; Jonsson 2009; Croidieu & Monin 2009), circumventing the goal of thoroughly understanding the diffusion of an innovation as an instance of technical and social change that goes beyond the attributes of the innovation. Thus, by terming the spread of corporate research an innovation, it shall by no means be assumed that the actors involved in its creation and diffusion perceived it as such. As historical cases show, there was an abundance of testing and control laboratories long before the first research laboratory was established. Additionally, one may also ask what exactly was diffusing: the physical object "laboratory", the floor plans of laboratories, the practice of doing research within a corporation, or an even more hazy idea of the benefits of science and research – with the diffusion of immaterial practices or ideas being heavily dependent upon their material means of travel, the profile their diffusion takes often depending on who arrives first, "the merchant or the missionary" (Katz 1999, 151; cf. Czarniawska & Joerges 1996, 36; Rogers 2003, 13f). This issue is further complicated when asking whether it was in fact the same practice spreading through the field of interest, with the variety of forms that industrial research laboratories took, and past research detailing the importance of partial diffusion and re-invention throughout the diffusion process (Rogers 2003, 180ff; Alasuutari 2015). Since the detailed accounts of the R&D pioneers convincingly illustrate the different ways that corporate research was undertaken, the item diffusing here is the *practice* of establishing an industrial research laboratory as a new organizational entity, whatever form it may take and wherever it may be located within the organizational structure. For our purposes, we will regard the industrial research laboratory as an *organizational novelty* that became an innovation once it was widely accepted and incorporated in large firms.

As these deliberations upon one of the key terms of diffusion studies illustrate, centering our explanation for the successful spread of the corporate lab around it being an innovation for actors of its time would obscure many of the factors involved in its diffusion (Djelic 2008, 545), namely contextual variables that are not only outlined theoretically in diffusion research but also hinted at in the existing histories of various early laboratories. Ideas, practices, and things obviously do not spread in a vacuum. Rather, the answer to why some ideas spread far and wide, while others disappear, can be found in the various local and temporal contexts these ideas move through:

> A fundamental element in adoption theory is recognition that innovations are not independent of their environmental context but that they rather evolve in a specific ecological and cultural context and that their successful transfer depends on their suitability to the new environments they enter during diffusion ....
>
> (Wejnert 2002, 310)

What Wejnert calls context can be found under various designations in diffusion studies, such as environment or social system, which in Roger's analytical scheme means "a set of interrelated units that are engaged in joint problem solving to accomplish a common goal" (Rogers 2003, 23), with its units variously being individuals, organizations or societal subsystems, and their interrelations forming a social structure that yields points of contact and communication. As Palloni notes, a proper explanation of a diffusion process needs an understanding of the structures involved: "Some of the best original work on diffusion processes emphasizes that social diffusion is an analytically sterile construct if not cast against a social structure ..." (Palloni 2001, 73). But as Strang and Soule note, merely accounting for social structure is not enough to understand why the industrial research laboratory emerged and diffused: "Structural opportunities for meaningful contact cannot tell us what sorts of practices are likely to diffuse, and such opportunities may lead to conflict or boundary formation as well as to diffusion" (Strang & Soule 1998, 276). This indicates that there are more than just patterns of interaction that influence successful diffusion.

In Roger's categorization of the four elements of diffusion, the aforementioned *social system* also contains system norms as established behavioral patterns, that tell system members how to act – in the case of diffusion whether to adopt or reject (Rogers 2003, 26). Wejnert (2002, 312ff) refers to belief systems or socialization, whereas Katz et al. (1963, 249ff) speak of value systems, and Palloni (2001, 73) adding "repertoires of feasible behaviors and preferences" that guide action within the social structure. Thus, many ways that social structure has been enriched by cultural variables in diffusion theorizing accentuate its importance in the spread of ideas, practices, and things, even though a common theoretical and methodical approach is lacking. This emphasizes, though, what matters is not only the innovation itself and the social structure that influences who hears about it, who has the means to adopt it, and their ability to influence others, but also the meanings assigned to (and transformed by) the ideas, practices, and things (and adopters) in the process of diffusion on the one hand, and the way those ideas manage to fit with these cultural attributes of the social structure on the other hand. Needless to say, diffusion will vary according to shifting spatial and temporal factors, as ideas will be interpreted differently in different places and at different times. The idea of what a laboratory is and looks like today evokes distinctly different images than one in the United States of the Progressive Era, and the practice of establishing industrial research laboratories looked decisively different in 1900 and, say, the 1980s, when the thought of investment into uncertain scientific innovation was challenged and ultimately delegitimized (cf. Mowery & Teece 1996; Carlson 2007). Similarly, the earliest instances of the establishment of industrial research laboratories can be found in the German dyestuffs industry, between 1877 and 1882, illustrating a different interplay of economic, political, and cultural factors than in the US chemical and electrical industries (Homburg 1992). This brings me to the core argument of this book: the meanings assigned to an innovation, and the larger, surrounding system of meaning that inform action, influence the diffusion of ideas, practices, and things, and need to be taken into account in order to explain diffusion mechanisms and the unique patterns diffusion processes exhibit.

Translating Ideas to Make Them Fit

Of course, the argument elaborated previously has been made before and put to test in a variety of ways both within diffusion studies – often to measure the "compatibility", "fit", or "appropriateness" of an innovation and the surrounding social system – and in other branches of sociology. What can be learned from these efforts within diffusion studies, and what new perspectives shall be added here?

As the seminal theorist of diffusion, Rogers emphasizes the necessity of fit in his discussion of compatibility as "the degree to which an innovation is perceived as consistent with the existing values, past experiences, and needs of potential adopters" (Rogers 2003, 240), with higher compatibility decreasing uncertainty and in turn increasing the tendency of adoption. He further distinguishes compatibility with system-level values and beliefs, as well as previously introduced ideas. Ultimately, although, according to Rogers, relative advantage proved to be the strongest predictor for the rate of adoption (i.e. how an innovation spread) rather than compatibility (ibid., 233). Croidieu and Monin (2009) put this to the test in their analysis of the (non-)diffusion of entrepreneurial innovations in wine making, finding that effective practices can fail to diffuse if deemed inappropriate, and noting how "potential adopters assessed appropriateness first, and then effectiveness" (ibid., 320). They further extend the notion of compatibility by underscoring how fit is always locally and temporally dependent. In a similar vein, Soule (1999) shows how the protest practice of building shantytowns on college campuses, although ineffective, diffused due to the appearances of appropriateness, thus being compatible. In his analysis of the evolution of the hostile takeover in businesses in the United States between 1960 and 1985, Hirsch showed how language associated with such takeovers changed to more benign terms over time, correlating "cultural symbols with structural movements" (Hirsch 1986, 814). Hirsch illustrates how ideas, practices, and things that may not be compatible with the larger cultural variables of a given social system can be *made* compatible, by building bridges toward that, which is already known and widely accepted – in this case, shared terms, metaphors, and narratives.

These empirical, case-based findings are further underscored by Strang and Soule's review of diffusion research (1998), where the authors discuss what they call *cultural bases of diffusion*. They find that "[j]ointly, the argument is that practices diffuse as they are rendered salient, familiar, and compelling" (ibid., 276). Fit or compatibility is indeed essential, and those practices that "accord with cultural understandings of appropriate and effective action tend to diffuse more quickly than those that do not" (ibid., 278; cf. Wejnert 2002, 313). As it turns out, legitimacy and perception of appropriateness are more important for central actors than for those who are marginalized at the fringes of a field (Strang & Soule 1998, 279; Leblebici et al. 1991). In addition, they find that practices themselves do not diffuse, but rather "theorized models and careful framings" of these practices do (Strang & Soule 1998, 277), further emphasizing the necessity of interpretive work in diffusion. This argument is extended by Strang and Meyer (1993), who outline how "theorizing", as the development of abstract categories and cause-effect relations (ibid., 492), aids diffusion.

In summary, *compatibility, fit*, or *appropriateness* with cultural attributes of the social system play an important role in the diffusion of an innovation. It is clear that these are not fixed categories – ideas, practices, and things can be made compatible and made to fit in the process of their diffusion. This active, constructive element is captured in the notion of translation.[1] Originally going back to the work of Michel Serres (1992), translation came to represent not only the linguistic meaning of translating words between languages but also the added dimension of the transformation of objects and material things. Widely used by scholars in the fold of Actor-Network Theory such as Latour and Callon, who emphasized the nature of translation as collective action (cf. Callon 1980, 1986; Latour 1986), the concept came to be incorporated into a body of new institutional scholarship often referred to as Scandinavian Institutionalism (Wæraas & Nielsen 2016). Out of an interest in global and local change processes, these researchers combined new institutional tenets with French constructivism. Even though no general theoretical lens and methodological approach exist, Scandinavian Institutionalist research often asks questions of local change processes, organizational practice, and fashions, while utilizing interpretive, qualitative approaches to data-gathering and analysis (cf. Czarniawska & Sevón 2003; Engwall 2003; Mica 2013).

But how do translation and diffusion differ? In the (original) formulation of the diffusion perspective, ideas, practices, and things spread unchanged, with innovators adopting the same item as laggards, no matter the time or place (cf. Mica 2013). Translation shifts the perspective to show who does the translating, what exactly is being translated, and why. The concept of translation thus adds several facets to this analysis. As Scandinavian Institutionalism's scholarship has illustrated, in the case of organizational forms it is not the one form that spreads: "What is being transferred from one setting to another is not an idea or a practice as such, but rather accounts and materializations of a certain idea or practice" (Sahlin & Wedlin 2008, 225). These generalized models are then incorporated by organizations as a whole, in parts, or merely as a label in the case of associating new practices with already legitimized concepts (Solli et al. 2005). The previously passive actor is no longer merely an "adopter", instead becoming an active "editor" (Sahlin & Wedlin 2008, 227ff). With the editing of ideas not being open-ended and completely free, translation is tied back to compatibility and fit as introduced previously: Editors have to obey certain editing rules (ibid., 225ff) for their translations to become successful. Editing rules are assumed to arise from the cultural attributes of the social system, which are vaguely described and only very hazily defined by notions such as "master ideas" (cf. Czarniawska & Joerges 1996, 36f), or the rationalistic values associated with the Western project of progress (Meyer 1996, 250). Yet it remains rather unclear and undertheorized how exactly culturally constructed meaning, collectively shared by members of a social system, influences the translation process – answers are still lacking both in diffusion research, as well as in the translation literature that builds on organizational institutionalism's core tenets.

The idea that there is more to a social system or any given society than patterns of interaction and communication is, of course, not unique to diffusion studies. Specifically, the conception of an interdependency between cultural variables and the material world could be seen as one of sociology's core interests and can be found in various formulations in macrosociological works. Evidently, the idea that the institutional-material order needs to be connected to an ideational layer, be it called societal semantics or a social system's given culture, holds merit. Yet there is a pressing need to reach a more accurate definition of what that elusive cultural element of a social system is, and how it goes about constructing and lending meaning to reality. Similarly, the translation literature hailing from an institutionalist background that updates diffusion scholarship in regard to fit and compatibility, especially through emphasis on how the cultural variables condition translation via editing rules, lacks clear concepts to theoretically and methodically capture these relationships on their own. These shortcomings shall be remedied in this book, by extending the translation framework of diffusion studies with a clear theoretical strategy of delineating a social system's cultural attributes, as found in organizational institutionalism's concept of the organizational field. Finally, by understanding the cultural attributes as discourse, a methodical approach is developed that captures most of the dimensions of a field's discourse.

Idea Translation in Organizational Fields

Since the diffusion phenomenon of interest here resulted in organizational change, it makes sense to turn toward organization theory for further elaborations on how organizational change processes operate and what conditions organizational behavior. A few words on organizational institutionalism[2] are necessary, to underline how some of its core tenets enrich the theoretical framework of diffusion studies and to sharpen the thrust of the analysis into the origins of early research laboratories.

The question of what makes organizations act the way they do is front and center in this line of thought. In the 1970s, organizations were "largely portrayed ... as agentic actors responding to situational circumstances" (Greenwood et al. 2008, 3) by frameworks such as resource-dependence theory, ecological theory, and behavioral theories. From these perspectives, the executives of organizations decided – more or less rationally – how to react to their environments, which were defined as their respective markets. These assumptions were challenged by the foundational texts of organizational institutionalism (Meyer & Rowan 1977; Zucker 1977; DiMaggio & Powell 1983; cf. Greenwood et al. 2008, 3ff). With regard to institutional theory, it can be argued that organizational institutionalism arose primarily from challenging the distinction between rationality and culture (Hall & Taylor 1996, 946; Nee 1998). While other institutionalisms, such as historical institutionalism, focus on power struggles as mediated by institutions present in political and economic structures, or rational choice institutionalism, which assumes rational actors whose collective action problems are solved by institutions (Hall & Taylor 1996, 946; Campbell 1998), organizational institutionalism broke with a focus on power and assumptions of rationality: "Anyone who has waited at a traffic light when no-one else was around, however, has to admit that there are dimensions to the relationship between institutions and actions that may not be highly instrumental or well-modeled by rational choice theories" (Hall & Taylor 1996, 951). Instead, new institutionalist scholars point out how agency is always dependent on a cultural repertoire of possible, legitimate actions (cf. Green & Li 2011, 1667) – a repertoire they call institutions. In the end, the goal of such an analysis shifts toward understanding how institutions condition organizational behavior not only in terms of "dos and don'ts", for example, by imposing sanctions, but also by highlighting the ways in which the possibility of action and the organization itself are rendered possible in the first place. This aids in generating sensitivity to the contingency of organizational decision-making beyond assumptions of managerial strategy and rationality on the part of corporate leadership, most of whom required a lot of convincing to actually establish laboratories. It also underlines the need to probe the institutions involved – those elusive cultural variables that will be analyzed using the concept and methods of discourse – that fill concepts such as the laboratory, research, or corporate innovation with meaning.

Of Organizational Fields and Institutions

As the first step toward circumscribing the reality that chemical corporations found themselves in at the turn of the twentieth century when the first laboratories were established, some boundaries need to be drawn since it would be foolish to attempt to include the totality of the social world in this analysis. I expanded upon the diffusion framework by using an institutionalist lens because the organizational field supplies a perspective that can offer ways of capturing an organization's environment beyond overly simplistic notions such as sector or network. Moreover, it also enables the drawing of clearer lines of demarcation around the early corporations that were establishing research laboratories, and more precisely than assuming they are in a mutually shared social system whose cultural variables condition their decision-making. With the organizational field, organizational institutionalism offers a well-developed vocabulary to account for just these factors.

Organizational fields were introduced by DiMaggio and Powell (1983) in one of the seminal contributions of organizational institutionalism:

> By organizational field, we mean those organizations that, in the aggregate, constitute a recognized area of institutional life: key suppliers, resource and product consumers, regulatory agencies, and other organizations that produce similar services or products.
>
> (ibid., 148)

Several of the features introduced here are key in outlining organizational environments. As the authors note, the existence of a field depends upon their institutional definition or "structuration", consisting of the interaction between organizations, patterns of domination or coalition, increased information loads, as well as a mutual awareness between field members (ibid.). This means that for an organizational field to exist, that is, to be recognized, it is necessary for organizations to interact in various ways and to be aware of these interactions. Scott further elaborates on this by noting how organizations within a field interact more with each other than with organizations outside the field, while sharing a "meaning system", as institutionally legitimized guides for action and reality perception (Scott 1995, 56; cf. Wooten & Hoffman 2008).

DiMaggio and Powell's concept, and its later extensions by other authors, allows for capturing organizations' environments in a way that lies transverse to other such attempts. Instead of using preconceived notions of industry (such as "the chemical industry") or sectors organized around common technologies or markets, and instead of letting competition be the structuring force, organizational fields account for the "totality of relevant actors" (DiMaggio & Powell 1983, 148), replacing geographical boundaries with cultural ones (cf. Scott 1994, 206).

When looking at the organizational field of those firms that have established industrial research laboratories, the focus shifts beyond just firms having a laboratory. Now suppliers of raw materials and consumers of end product are included, while the agencies imposing regulations upon these corporations need to be accounted for as well. Also, the notion of field goes even further, since those organizations supplying staff and expertise – such as universities and colleges, scientific and professional societies, or independent laboratories – form a recognized part of institutional life as well.

A field's boundaries are not fixed but permeable – both analytically and empirically. Analytically, since, as Sahlin-Andersson remarks, fields are "demarcated by the eye of the observer" (Sahlin-Andersson 1996, 73f), and their borders are dependent upon research interest – a study based on the same set of corporations but asking the question of, say, tariff legislation impacts, will draw wholly different boundaries. Fields are permeable empirically, as Davis and Marquis point out, as the "players that populate fields and the nature of their play can change over time" (Davis & Marquis 2005, 337). Their review of field-based research emphasizes how the notion of organizational field can be especially helpful when industries transform and their boundaries shift and change (ibid.).

Later work using organizational fields as units of analysis enriched the approach by composing fields as not simply the transactions between organizations, reacting to criticism of organizational institutionalism's difficulty in explaining endogenous change, that is, change going beyond exogenous shocks such as economic crises (cf. Walgenbach & Meyer 2008, 73ff). Hoffman (1999) in particular extended the notion of field by centering organizational fields around issues, as "centers of debates" (ibid., 351) that bring together a variety of field constituents having wildly different stakes in the discussion, thereby making fields much more like arenas of struggle and change than in the original conception by DiMaggio and Powell (cf. Wooten & Hoffman 2008, 134f). In the end, understanding the field of R&D laboratories as being organized around the issue of corporate innovation aids in capturing all possible relevant actors and themes. Of course, the contents and structure of a field have to be investigated empirically, not a priori (DiMaggio & Powell 1983, 148), making the charting of its elements and boundaries important parts of the analysis of the diffusion process at hand.

Organizational institutionalism posits that within an organizational field there are not only organizations that interact with each other in certain, patterned ways, thereby yielding a social structure, but also institutions that essentially form the cultural variables identified by diffusion research as playing a large part in successful diffusion. Yet, looking back at forty years of theorizing within organizational institutionalism, it may seem surprising how its key concept, the institution, remained ill-defined and fuzzy in early contributions. Later work remedied these shortcomings somewhat allowing Greenwood et al. to formulate an apt summary of what an institution is:

> … more-or-less taken-for-granted repetitive social behaviour that is underpinned by normative systems and cognitive understandings that give meaning to social exchange and thus enable self-reproducing social order.
>
> (Greenwood et al. 2008, 4f)

With this broad definition, institutions do not only include procedures and norms but also cognitive scripts and moral templates, providing frames of reference for meaningful action (Hall & Taylor 1996, 947). In contrast to the conception of institutions as explicit rules or conventions, as found in historical institutionalism, for sociologists, the world is full of institutions, ranging from "handshakes to marriages to strategic-planning departments" (DiMaggio & Powell 1991, 9). Institutions can thus be found on various levels. On the individual level, where they may take the form of handshakes; on the organizational, as structural elements such as equal opportunity commissioners or practices such as accounting; on a field level reflected in status hierarchies between occupations; or even on the societal level, such as the idea of national sovereignty (Greenwood et al. 2008, 5).

Institutions are purported to possess a certain durability – the power to stay and resist change or attempts at modification – aiding in the reproduction of social behavior and reality definitions, that is, situations are understood in the same, shared way, and action follows along the same, expected lines. Moreover, out of a rejection of any assumptions of rational-actor models, in this perspective institutions are no longer the product of conscious design or even purposive, instrumental planning, as other schools of institutionalism would argue (DiMaggio & Powell 1991, 8). This is where the oft-mentioned taken-for-grantedness comes in – "We've always done it like this!" – as institutions exist in shared bodies of knowledge, providing typologies for actors and scripts for action (Berger & Luckmann 1967; Zucker 1977; Barley & Tolbert 1997). But taken-for-grantedness goes further than being a result of the sedimentation of repeated action, by having institutions define reality: "Institutions influence behaviour not simply by specifying what one should do but also by specifying what one can imagine oneself doing in a given context" (Hall & Taylor 1996, 948). As a result, organizational actorhood and the concept of rationality itself are a consequence of the institutional (and with it, discursive) setup of the social world.

From Institutions to Isomorphism

The organizational field aids in outlining boundaries and gives a clear guide for a research strategy, while underscoring how important it is to go beyond the confining laboratory walls of the R&D pioneers that will be discussed in detail later, and toward their organizational environment. Supplanting the previously used variable of social system with the organizational field, an organization's surroundings are elevated from passive to active by highlighting how "environments operate to enact [socially construct] organizations" (Scott 1994, 208). Instead of scrutinizing individual laboratories and the internal, individual decisions that led to their establishment and continued existence, this analysis turns outward – because answering the question of why chemical firms around 1900 chose to establish laboratories within their walls means understanding these firms' organizational reality as being made up of not only competitors, suppliers, universities, and inventors but also by notions – that is, institutions – of proper corporate strategy, what a laboratory is, and so on. While in historical scholarship (e.g. Carlson 1997; see also Chapter 3) the singular, heroic actor is often put front and center, making the seemingly large shift in corporate innovation solely the product of individual action, here, the explanatory mechanism is shifted from an individual level toward a field level. Empirical scholarship in this tradition has offered various ways of operationalizing the organizational field in different settings (e.g. Mody 2011, 2016).

By adding the concept of institution, it has also developed a sensitivity for the impact of field-level mechanisms such as isomorphism. Institutional isomorphism, as the homogenization of organizational structures, arises due to organizations reacting (or being made to react) in similar ways to environmental conditions, that is, what is happening in the organizational field (DiMaggio & Powell 1983, 149). Generally, three mechanisms producing isomorphic change are distinguished analytically: coercive isomorphism, the forcing of organizations to change due to formal or informal rules; mimetic isomorphism as imitation due to uncertainty; and normative isomorphism due to increasing professionalization in modern societies, with professional values creating ideas of what is proper (ibid., 150ff; Mizruchi & Fein 1999; Walgenbach & Meyer 2008, 35ff; Boxenbaum & Jonsson 2008). In this way, organizations, in reacting to other organizations who summarily form an organizational field, change according to field-level institutions that circumscribe what is rational, proper, or necessary. But organizational institutionalism's insights on isomorphism serve not only to show the effect of a field's institutions upon organizational behavior but also to illustrate a key goal of said behavior as an answer to why organizations are influenced by the field at all: the striving for legitimacy (cf. Deephouse 1996).

Previous scholarship shows how isomorphism and diffusion processes are related, such as Tolbert and Zucker (1983), who illustrated how the reasons for adoption in the diffusion of civil service reforms changed during the diffusion process, with later adoption being subject to isomorphic pressures, whereas Greenwood et al. (2008, 11f) noted how the mere diffusion of any idea or practice does not have to signify isomorphism since the motivations for adoption can go beyond concerns for legitimacy. With the entanglement of diffusion and isomorphism proving to be more complex than one might initially assume, far be it from me to simply posit field-level effects resulting from the organizational quest for legitimacy as the answer to successful diffusion. Instead, this analysis aims for a more subtle illustration and understanding of isomorphism.

In focus here are normative and mimetic isomorphism, since the third kind, coercive isomorphism, is usually characterized as coercive measures applied by the state in the form of new laws and regulations. As the history of the early research laboratories shows, regulation and state-level actors played no role in their emergence beyond antitrust legislation. But, as shown later, understanding the birth of the R&D laboratory as a simple and direct consequence of business strategy shaped by legislation, misses key elements in its formation. Furthermore, coercive isomorphism in such an organizational field can, historically, only be identified at a later date, arguably with the advent of the national science landscape that took shape around World War II.

Mimetic isomorphism is a result of uncertainty, such as in situations of ambiguous goals or poorly understood technologies, leading to organizations modeling themselves "after similar organizations in their field that they perceive to be more legitimate or successful" (DiMaggio & Powell 1983, 152). Of course, this begets the question of who – or what – is perceived as successful, and whether said success or legitimacy is an organizational attribute, or instead constructed collectively. This holds true for normative isomorphism as well, which is a product of increasing professionalization. The creation of a formal knowledge base, as well as professional networks, lead to the diffusion of professional norms that even go beyond the attempts of the profession to define the boundaries and content of their work, and toward common styles of dress, behavior, and vocabularies (ibid., 153). How these norms arise and are communicated is itself a process of construction of what it means to be a professional, what the role entails, and so on. Simply ascribing both processes of construction – who is successful and what does it mean to be a professional – to the institutions of a field would fall short of the analytic goals of this study, which is why institutional isomorphism, as a field-level mechanism guiding the behavior of organizations, shall be understood as mediated by discourse. In short, isomorphism can be understood as the mechanism that translates the reality-setting aspect of institutions into the concrete demands made upon organizations, namely the need to attain legitimacy in the eyes of others. What legitimacy entails – what is perceived as legitimate – is dependent upon the organizational field's institutions, and, as I argue, constructed collectively, which can be analyzed as the discursive construction of institutions.

Before I proceed to what discourse is, it is imperative to outline where the link between institution and discourse lies, and why such an understanding is pursued here. On the one hand, adding the notion of institution to the theoretical ensemble has helped to clarify what the cultural attributes of social systems, as identified by diffusion research, are. These cultural attributes hone the understanding of what happens in organizational fields as well as just how much of organizational behavior can be ascribed not to rationality, but to reactions to man-made institutions that have become reified and taken for granted via institutional effects such as isomorphism. But the fundamental and subsequently most important aspect of institutions is how they go beyond mere rules and constraints on action and toward the definitions of reality itself – constructing what is possible and conceivable in the first place. From this perspective, in order to understand the successful diffusion of the R&D laboratory as a new organizational entity, simply looking into what concrete pressures and constraints influenced the decision-making is not enough. Rather, an understanding must be determined by looking at how research was institutionally mediated, what a laboratory was to organizations of the time, who the professional scientist claimed to be, and how these chemical corporations perceived their institutionally delineated fields of action. This brings me to the other hand: Even though the concept of institution aids in better accounting for organizational behavior and reality, reconstructing precisely which institutions played a role, where they came from, and how they were transformed in the first decades of the twentieth century is a task for which organizational institutionalism offers little help. Even though decades of scholarship was conducted on the ways that institutionalization and institutions work (as found in Scott's model of the three pillars supporting the institutional order, cf. Scott 1995), and on institutional change (e.g. Leblebici et al. 1991; Greenwood & Hinings 1996; Tolbert et al. 2011), there is little guidance on how to proceed methodically, due to the theoretical vagueness of the institution itself. Hence, this analysis sets out to contribute to the ongoing debates in institutionalism, and to develop a new way of operationalizing institutions and making the more-or-less subtle ways that institutions delineate reality while causing field-level effects in isomorphism visible for the case of the industrial research laboratory. To do this, I turn to a new concept: discourse.

Making the Case for Discourse

So far, the argument for how best to trace the diffusion of the industrial research laboratory has mostly remained within conceptual lines. Now it is time to get closer to the subject matter by proposing a way of analyzing an organizational field's institutions that manifest in isomorphic pressures – by way of discourse analysis. In organizational institutionalism, a multitude of methodologies has been used for scrutinizing fields and institutions (cf. Greenwood et al. 2008). Combining the study of organizations with conceptions of the nature of language use and reality is a relatively new way of attacking the problem of institutional change (cf. Phillips & Malhotra 2008). Even though no clearly defined procedure exists, mainly due to the many varieties of discourse, the concept of discourse has found fertile ground in organization theory in the last twenty years, producing a body of literature referred to as organizational discourse analysis (ODA) (Alvesson & Kärreman 2000b; Hardy 2001; Phillips & Malhotra 2008). But discourse has also found use in institutional scholarship called discursive institutionalism (cf. Schmidt 2008; Alasuutari 2015) and expressing similar ideas as ODA. However, it is less focused on organizations, instead of diving into the relationship between discourse and institutions.

What Discourse Is, and What It Is Not

Before delving into the details of discourse analysis – which always incorporates theoretical ideas about what discourse is – and what it does, the relationship between institutional and discourse theory needs to be clarified. Why adding another concept to the theoretical model of ideas being translated throughout an organizational field? Does *discourse* add what *institution* currently lacks?

The key argument for how adding discourse enhances the understanding of institutionalization can be found in Phillips et al. (2004). The authors emphasize that organizational institutionalism focuses too much on institutions and too little on the actual actions involved in the process of institutionalization, ignoring the cognitive foundations of institutions (cf. Phillips & Malhotra 2008). To remedy this, Phillips et al. (2004, 638) argue that institutions are, essentially, text-based entities. They illustrate the relationship of action and text with the advent of the multidivisional form adopted by US corporations. Instead of witnessing the actual organizational restructuring "in action", executives read about it in books and magazines, heard stories about it, saw organizational charts, and consequently acted upon them (ibid., 639). In this way, the multidivisional form came to be an institution (as a certain form of organizing), not because of sedimentary, repeated action, but because of action that was materialized in texts and thereby made iterable, transferable, and repeatable. What becomes relevant, then, is not "simple imitation of an action by immediate observers" (ibid.), but the creation of supporting texts that make up and are influenced by discourse. From the discursive perspective, institutions are understood as enacted discourse, shaping action and, in turn, text-production, shifting focus from what is "doable" within an institutional order, toward what is "sayable", that is, what can be put into text. This change of perspective is the key contribution of ODA to an institutionalist research program, offering the possibility to understand institutions that are made up of collections of texts. Now the analysis of institutionalization and organizational change becomes a question of discursive effects, that is, which texts that are produced stick, and why (ibid., 640).

Bringing text, and with it language, to the center of institutional analysis and emphasizing the constructive effects of language itself, are a consequence of the linguistic turn in the social sciences in general, and in organizational studies in particular (Phillips & Hardy 2002, 12f; Deetz 2003). This breaks with the assumption that language is a representation of reality, and posits against "a conventional view of language as a transparent medium for the transport of meaning" (Alvesson & Kärreman 2000a, 141). Instead, language is now seen as ambiguous, context-dependent, repressing hidden meanings, and built upon unsure foundations in general (ibid.). In rejecting the so-called language-as-mirror logic (ibid., 138f), the constructive effects of language come into focus: "It constructs reality in the sense that every instance of language use is to some extent arbitrary and produces a particular version of what is it [*sic*] supposed to represent" (ibid., 142). If the language does not represent reality but constructs it in the first place, then what becomes important is to analyze how this happens: how exactly a certain version of reality is made by language, manifested in texts, who is involved, and how competing realities are excluded. This is explicitly pointed out by Deetz, who laments that the linguistic turn in organizational analysis often shifted attention from language being the mirror of nature and toward merely analyzing said mirror as an object, scrutinizing "texts and talking rather than looking through discourse to see the specific ways the world is produced" (Deetz 2003, 425). This involves analyzing how actors draw on and transform discourses that constitute a specific social reality, as well as analyzing how meaning is negotiated, made intelligible, or ruled out (Grant & Marshak 2011, 207f).

What is the relationship, then, between discourse and institutions? Institutions are made up of action condensed in texts, that is, in language, and language is not merely a representation of said action, but always puts the action in a certain context, and constructs a certain reality in the process. The bodies of texts making up an institution are called discourse. Consequently, discourse is not an intermediary concept, lying "between" the actors in the organizational field and the institutions, but rather a way of conceptualizing what institutions are made up of – texts – and how they exert their pressures and reality-building effects: by delimiting what is sayable, and, at its core, determining what things *are* in the first place. Thus, discourse analysis offers a sensible framework, especially since the act of translation is often a textual practice, and second, a pragmatic avenue of inquiry had to be chosen in order to analyze a diffusion process that took place more than one hundred years in the past.

Similar to discourse analysis's many meanings (Alvesson & Kärreman 2000b, 1127), discourse itself is a concept that offers a plethora of definitions, often differing according to each theoretical perspective and intended use. The approach to discourse developed here ultimately dates back to Foucault's historical analyses (1981), in which he advanced the idea that language constructs the social world, giving rise to certain subjectivities and social practices (cf. Alvesson & Kärreman 2000b, 1127f). Building upon this foundation, ODA takes inspiration from critical discourse analysis (CDA) in the way discourse and its effects are defined (cf. Fairclough 1992; Wodak & Meyer 2001).[3] Even though ODA does not share CDA's strong focus on relations of power and how control is exerted discursively, the inspirations taken from CDA's approach have served to include the social context, going beyond the discursive realm and managing to mold discourse into a multidimensional concept.

ODA defines discourse as "an interrelated set of texts, and the practices of their production, dissemination, and reception, that brings an object into being" (Phillips & Hardy 2002, 3). This sentence contains all relevant ingredients encapsulating what discourse is about. First, discourse is made up of multiple texts. Coming from a perspective heavily influenced by French post-structuralism, a broad view on what "text" means is taken. Hence, text not only refers to the written word but also contains speech acts, pictures, artifacts, or symbols (Phillips & Hardy 2002, 4; Grant & Marshak 2011, 208). Yet, to become texts, anything written or said must be inscribed, taking on a material form and making it accessible and transferable beyond the immediate situation (Phillips et al. 2004, 636; cf. Fairclough 1992; Chalaby 1996). But a text as the basic discursive unit is not meaningful itself, instead discourse refers to interrelated, structured sets of texts. This means that texts need to be seen, heard, or read to become meaningful, and that they need to relate to each other in some way – be it by describing a shared topic, drawing upon the same discourses, explicitly referencing each other, or their conditions of production and reception (Phillips et al. 2004, 636).[4] Thus, discourses are neither random collections of texts nor diary entries or other invisible texts.

Second, texts always need to be situated within their context: "Discourse is not just ideas or 'text' (what is said) but also context (where, when, how, and why it was said)" (Schmidt 2008, 305). Context comes into play in several forms. Especially in regard to practices of production, dissemination, and reception, texts need to be situated within their temporal, historical, and social contexts (Grant & Marshak 2011, 208; cf. Chalaby 1996, 688ff; Van Dijk 1997). Fairclough (1992, 71ff; 73, fig. 3.1) elaborates on this, noting how both discursive and social practices influence the production and reception of texts. Discursive context means not only the necessity of analyzing how texts relate to discourses and each other (cf. Wodak 2001) but also how discourses influence the form and content texts can take, by shaping "what can be said and who can say it" (Grant & Marshak 2011, 208). Obviously, there is a multitude of actors within an organizational field, structured by both status and communication. Thus, it matters who says what, that is, produces certain texts, since actors have different means of making themselves heard. Phillips et al. (2004, 643) characterize this as positions warranting "voice", legitimate speakers who can make their texts "stick" via resources, coercive means, or by being centrally positioned in an organizational field. Examples include experts who are perceived to have authority over a certain topic, such as a psychologist on the topic of sanity (cf. Munir & Phillips 2005, 1668), which gives them "voice". Large, wealthy corporations have the means to make texts stick due to the power they exert over suppliers and competitors while being highly visible and possible targets for mimesis. Furthermore, texts need not only be related to actors and their respective relationships, but also to events shaping what is being talked about, as well as field-level institutions conditioning the ways texts are produced. In this way, a collection of texts making up a discourse always needs to be understood within this three-dimensional conception of discourse, that is, within a textual, a discursive, and a social dimension (Fairclough 1992, 73; Phillips & Hardy 2002, 4).

Third, discourses bring objects into being. Due to discourse theory's inherent assumptions about the nature of language, discourses are not of interest because they are somewhat structured sets of texts, produced by various actors in differing situations; they are interesting due to their constructive effects – discourses are not just a certain thing, they *do* something as well. The concise definition from Phillips and Hardy only mentions objects, but discourse also constructs ideas and identities (Phillips & Hardy 2002, 4; Grant & Marshak 2011, 208; cf. Fairclough 1992), in the end constructing reality itself, "giving it meanings that generate particular experiences and practices" (Phillips et al. 2004, 636). Hardy et al. (2005, 60) give several examples of the constructive effects of discourse, such as managerial discourse creating ideas like "strategic planning" shaping business practices. Hardy argues that concepts such as "the organization" are discursively shaped, "by fixing their identity so that it becomes possible to talk about them *as if they were* naturally existing social entities" (Hardy 2001, 28, emphasis in original).

Even material objects are made meaningful through discourse itself. The goal of discourse analysis, then, is to make visible the various ways in which reality is produced. Ideas, concepts, and things are constructed and made meaningful in practices of text production and are translated according to discursive, that is, institutional, editing rules, ruling in and out not only action but also speech.

Fourth, and last, discourse cannot be studied directly. Instead, "they can only be explored examining the texts that constitute them" (Phillips et al. 2004, 636). The text as a basic discursive unit also serves as its material manifestation (Chalaby 1996, 688). Heracleous and Barrett circumscribe the relationship between text and discourse as akin to that of action and structure: "Just as the structural properties of social systems are, according to Giddens, instantiated as social practices, so the structural properties of discourse are instantiated in daily communicative actions" (Heracleous & Barrett 2001, 758). This means that, while discourse is manifested in individual texts, it also exists beyond these texts, with each text being a trace of one or several discourses. The methodological implication follows that analyzing discourse means systematically studying multiple texts to understand the ways in which discourses construct meaning and, accordingly, reality.

After defining precisely what is meant by discourse, some cautious notes are necessary to limit the notion and show when discursive effects do not apply – what discourse is not. Alvesson and Kärreman (2000a, 2000b, 2011a, 2011b) caution against a definition of discourse that is either too vague or too muscular. Reviewing recent work in ODA, they lament that many authors ground their works on a very loosely defined concept of discourse: "'blow the concept up, use it ambiguously to say everything and nothing' sometimes appears to be the guideline" (Alvesson & Kärreman 2011a, 1195). This deteriorates discourse into a re-labeling of existing concepts such as culture and ideology, at times "capturing almost the entire sphere of the social and cultural" (ibid.). Thankfully, such a pitfall can be avoided by clearly defining the concept, and by locating it within clear theoretical boundaries – in my case discourse as textual expressions of institutions. In this way, discourse as a theoretical concept (and as a method) can be combined with the conceptual arsenal of organizational institutionalism and several of its well-developed notions such as the organizational field. Here, discourse is not "everything social", but sets of textual materialization of actions that are shaped by discursively constructed institutions. Furthermore, the fact that discourses *do* things (construct reality) needs to be investigated, and not posited as a starting point of an analysis that then goes on to explain everything by way of discourse (Alvesson & Kärreman 2011a, 1199; Iedema 2011). Alvesson and Kärreman advise against this by emphasizing how taking a discourse's power for granted may lead to an "inclination to overplay the discourse card (explaining everything) rather than to use the concept to explore (the limitations of) its constitutive powers" (Alvesson & Kärreman 2011b, 1133). Elucidating the junctures of discourse and practice, that is, text-production in an area of conflict over what is sayable and what is not, should be the goal of any serious work utilizing the concept of discourse.

In the end, it helps to ask "how" questions in order to avoid letting discourse become the be-all and end-all of sociological explanations. How did action captured in texts produced by organizations in the contested field of US chemistry contribute to any overarching discourses, for example, on corporate innovation? How and why did the ideas or objects constructed by these texts garner staying power? How was the idea of an industrial research laboratory – the "text" of interest to this study – translated not only by those spreading it from corporation to corporation but also by discursive institutions setting expectations and limits to the editing of an idea? These questions are reformulations of the original questions – how do ideas spread, and specifically how did the idea of corporate R&D laboratories spread – within the proposed theoretical model of idea translation in organizational fields. In the following, some operationalizations are necessary to figure out a good place to start finding answers.

Toward an Operationalization of Discourse

Similar to many definitions of discourse, many different approaches to actually analyzing discourse exist, informed by a theoretical perspective and methodical toolbox, leading to a large variety of research that calls itself discourse analysis, yet often with few common features (cf. Van Dijk 1997; Keller et al. 2010). Even within ODA, as a loose set of scholarship focused on the effects of discourse on organizations and organizing, there is no shared procedure with regard to operationalizing discourse and assembling a corpus of texts for analysis (Alvesson & Kärreman 2011b). This is due to the fuzziness of discourse as a concept, but also the many different ways of approaching textual data, ranging from genre to narrative analysis, and from linguistics to ethnography (Phillips & Hardy 2002, 9). Furthermore, the best way to analyze texts is often highly dependent upon the types of texts – be they archival records, interviews, novels, or cartoons – as well as the research question guiding the whole inquiry, and the nature of the organizational field in which the discourse of interest is materialized. That is to say, there is no clear sequence of discursive methodology to follow, and many of the decisions on how to analyze the discourse in question can only be answered once the organizational field and the nature of text and text-producing entities are known. As such, the goal of this chapter was to introduce discourse as a theoretical concept at the intersection of diffusion and institutionalization, and now it is to break the concept down in a way that is meaningful to the research question. The actual work of assembling a corpus and developing a schema for analysis can only be done after more is known about the history of the early R&D laboratories.

The first step toward an operationalization is to locate the question of the diffusion of the industrial research laboratory within the usual range of discourse analysis, as captured in the schema discussed by Phillips and Hardy (2002, 18). They propose that, at the beginning, researchers doing discourse analysis need to decide whether they are more interested in questions of power or social construction and whether they want to focus more on individual texts or the surrounding context. Studies putting the central emphasis on individual texts will often favor linguistic microanalyses of a few texts, whereas those focusing on context will highlight the way social and other conditions of production and reception influence texts. The difference between a constructivist and a critical approach boils down to whether the analysis is more interested in the constructive effects of discourse leading to the creation and reification of ideas and objects (Hardy 2004, 416), or in the use of power as made possible by language, as akin to Foucault's work.

Obviously, an interest in the discursive construction of reality lies at the heart of the questions at hand here, how the idea of the R&D laboratory was translated into varying contexts. Even though they could be framed as questions regarding power – the power of large corporations, the power of groups vying over epistemic authority – ultimately, the constructionist pole predominates, since in this diffusion process we are less interested in specific situations where power was exerted, and more in how an idea came to be materialized and changed due to its surrounding discursive context. Within this choice of constructivism over critique lies a preference for context over text, given both by research interest and theoretical framework. Research interest is geared toward understanding how ideas spread and become institutionalized, a process that can only be understood within its context. The theoretical framework of combining basic assumptions from diffusion research with concrete notions from organizational theory in order to get a clear delineation of a social system – an organizational field – and what is in it – institutions that are products of action-as-text – also offers the means to provide the three-dimensionality that Fairclough (1992, 73) demands. While microanalyses of select texts may offer clues of how the concept was translated, in the end, it is the overarching, grander discourses around the turn of the nineteenth century that are assumed to have influenced how the idea of a corporate laboratory was made resonant. These discourses need to be identified analytically.

Due to the constructive orientation, the way that social reality is generated by discourse is in focus. Following Fairclough (1992) and Phillips and Hardy (2002) to break construction further down, it is useful to distinguish between three types of entities constituted by discourse: *concepts*, *objects*, and *subject positions*. Concepts are "culturally and historically situated frames for understanding social reality – ideas, categories and theories through which we understand the world" (Maguire & Hardy 2006, 13). Munir and Phillips (2005, 1168) give the example of "endangered species" as a concept – the idea that an animal or a population of animals can be in danger due to poaching or environmental changes. Other examples would include managerial strategies, such as Total Quality Management, as ideas about how to lead and organize work and communicative relations between members of an organization. Thus, concepts reside "in the ideal" (ibid.) and refer to sensemaking and ways of making reality meaningful. Objects, in turn, are things that are made meaningful only through concepts. They have an "independent physical or material existence" (Maguire & Hardy 2006, 13), such as – with reference to the example of endangered species – a Snow Leopard. Far be it from discourse theory to negate a leopard's existence in a material way, but, by constructing it as endangered, the Snow Leopard becomes an object – the materialization of a concept. Of course, many different objectifications of leopards are possible, and while becoming an object may influence a Snow Leopard's reality (e.g. due to bans on hunting, etc.), its existence is in no way presupposed by discourse. Similarly, managerial practices are objectified in charts, plans, guidelines, or presentations. The main difference between concepts and objects is the material aspect of objects.

Lastly, subject positions are "locations in social space from which more or less well-defined agents produce certain kinds of texts in certain ways" (Munir & Phillips 2005, 1668). The example cited here is a psychiatrist certifying someone as insane – producing a special kind of text (a diagnosis), which is only possible due to the position inhabited by the psychiatrist. In a related vein, one needs a certain place in social structure – what sociology would call a role – to decide and communicate that Snow Leopards are endangered. Subject positions thus enable actors to produce texts in a legitimate way, while also being effects of discourse themselves, which means who is a legitimate speaker on what topic is discursively constructed in the first place by attributing authority or expertise.

These three features fit well within the framework of discursive institutionalization presented by Phillips et al. (2004) and the surrounding institutionalist ideas presented previously, offering positions for actors that have "voice" and are able to make texts "stick", without denying material reality and making everything discursive. Additionally, they serve as anchoring points for the two kinds of isomorphism of relevance here, with concepts and objects offering an avenue into what the organizational reality was made up of, and what could be perceived as legitimizing practices, for example, predominant concepts about business strategy, or materializations of such concepts in the ways organizational structures were realized. Subject positions can show how professional identity and its respective domain were constructed and subsequently carried into organizations in the first place. Moreover, capturing discourse along these lines offers clear guides for handling texts in individual analysis, while also making it possible to relate the findings to the larger framework of the translation of an idea and the subsequent production of institutions – How was the concept of the industrial research laboratory constructed, materialized in an object, and by whom?

The Gist of It

Let me now review the theoretical elements presented so far, in order to outline the next steps taken in analyzing the discourse surrounding early US industrial research laboratories. From cursory reading, we know that the practice of establishing research laboratories in US corporations started around 1900, mainly in the chemical and electrical industries. The organizational entity that came to be known as the R&D laboratory then spread rapidly throughout these and other industries. Thus, we have an instance of the diffusion of an innovation: a new element of the modern corporation made its way through many sectors of the American economy. From the long history of diffusion research, we know that it is not necessarily innovativeness, efficiency, or relative advantage that makes actors adopt a spreading idea. While the perceived attributes of an innovation – if it may even be called that – do certainly play a part in the process of diffusion, what came to be known as the cultural variables of the social system need not be missed. In other words, the surrounding social as well as cultural structures heavily condition successful diffusion: for ideas, practices, and things to travel successfully, they need to fit with established norms, values, and predominant cultural models. Furthermore, "fit" is not fixed, as later scholarship on translation emphasized – ideas are changed in the process of their diffusion, and are made to fit, or translated, into a variety of new contexts. Thus, in order to understand how the industrial research laboratory came to be translated successfully, the research question needs to be reformulated toward an analysis of the social system surrounding the corporations who established such laboratories.

In a second step, a vocabulary for delimiting the social system and understanding how it would and could affect organizations was needed, hence the turn to organizational institutionalism. With the organizational field, this perspective on organizations offers the ability to capture the corporations establishing laboratories beyond the crude or imprecise assemblages such as an industry or a sector. Instead, the field surrounding these corporations is composed of all relevant organizations, from suppliers to vendors, from universities to regulatory agencies, from testing laboratories to independent inventors; all centered around the issue of corporate innovation – how to improve existing products, to stay competitive? But organizational fields are more than a mere assembly of actors, they are also host to institutions. With the conception of the institution as cognitive scripts and normative understandings, resulting not only in repeated, taken-for-granted actions, but also in conditioning an understanding of agency and reality itself, organizational institutionalism offers clues as to what drives organizational decision-making in a field. Three types of isomorphism – coercive, normative, and mimetic – are the mechanisms in which institutional reality-setting manifests concretely in organizational practice. Organizational institutionalism thus offers tools for circumscribing the field around the early R&D laboratories without resorting to vague, undertheorized concepts that hamstring diffusion studies and other approaches utilizing similar understandings of the material and the ideational.

Of course, this formulation of the institution and its connection back to the field through the isomorphic mechanism is lacking in clarity, which is why, in a third step, discourse was introduced for analyzing the empirical expressions of a field's institutions that drive isomorphism. Since institutions arise from repeated action, and the aforementioned action is seldom observed, what is relevant, then, are recordings of these actions in texts, for example, in rules, plans, or reports. Discourses, as sets of interrelated texts, constitute reality and shape what is sayable and, subsequently, doable. Thus, acts of text-production within the organizational field need to be scrutinized in order to understand how the industrial research laboratory was talked about, how it was translated, and what exactly it was made fit to – what bridges of resonance were built. In the first attempt at operationalizing discourse, emphasis was put on what concepts were involved in legitimizing the laboratory and how the laboratory itself was conceptualized, how it was materialized in objects, and what subjects were involved and constructed in the process of its diffusion. With these three categories, isomorphism can be grasped more clearly, since they illustrate what – and who – is sayable and doable and subsequently perceived as legitimate, while the way these concepts, objects, and subject positions are constructed in individual discursive texts and made to fit shows the role of fit and appropriateness with existing institutions of an organizational field.

The individual steps of the research program follow from these conceptual developments. Chapter 3 will disassemble the existing history of R&D laboratories in the United States by charting their growth and then diving into the histories of the so-called research pioneers: General Electric, Du Pont, AT&T, and Eastman Kodak. The histories of their individual laboratories will serve to assemble the first collection of organizations, events, and individuals involved in the emergence and spread of research laboratories, as well as to scrutinize the theoretical hints that can be learned from the four pioneers. One key aspect from these stories is the role that the emerging idea of "science" plays in

American society of the time, with scientific disciplines gradually professionalizing and imprinting their promises and feats within the public consciousness, and with it – presumably – discourse. Thus, the focus is laid on what ideas of science and the scientist were present in the field, that is, what science means to organizations that establish a laboratory and how the institution of "science" impacts this process of organizational change. The assembled collection of organizations is utilized to further outline the organizational field in Chapter 4. It was chosen mainly to focus on chemical corporations for reasons that will be discussed later. Charting the organizational field involves analyzing US chemistry in academia, business, and politics. Armed with a clear delineation of the field and its relevant actors, events, and themes as compiled in Chapter 4, Chapter 5 will serve to further the operationalization of discourse by showing how the field's actors were transformed into speakers, how access to the texts they produced can be gathered, and how a strategy for analyzing these texts – and with it, the discourse – was formed. Chapter 6, then, will illustrate the analysis of the discourse on science in the organizational field surrounding the early corporate laboratories, spanning roughly thirty years. Breaking the discourse down into concepts, objects, and subject positions, the analysis will show which conceptions of science, research, the laboratory, the scientist, and the chemist predominated, and how they changed and were translated over time to fit new business contexts and societal challenges such as World War I. Lastly, Chapter 7 will offer a summation of results, bringing them back into the fold of diffusion theory to see how the story of the industrial research laboratory is enriched by an institutionalist, discourse-centric approach. Furthermore, the results will be discussed in the larger context of theory building, discourse analysis as a method, and historical and sociological research in general.

Notes

1 Alasuutari (2015, 179) remarks how there are multiple concepts tackling similar issues, such as creolization, glocalization, indigenization, or localization; but as translation is widely used in organizational scholarship, it shall be utilized here instead of these other notions.

2 This strand of organizational thinking today goes by many names: sociological institutionalism (Hall & Taylor 1996), new (or neo) institutionalism (Brinton & Nee 1998), or organizational institutionalism (Greenwood et al. 2008). In the following, the term organizational institutionalism will be used.

3 Of course, as is often the case in discourse analysis, CDA cannot be seen as one conjoint method of operationalizing discourse. Studies identified as CDA often share an interest in power struggles and the constructive effects of discourse, while varying in their approach to textual analysis (Machin & Mayr 2012; cf. also Phillips & Hardy 2002, 25f).

4 As Chalaby (1996, 689) remarks, in a sociological definition of discourse, texts become interrelated due to extra-discursive reasons (contrary to Foucault's conception), emphasizing the need to include external (social) conditions of production when speaking of interrelated texts, for example, by accounting for speaker or field as a possible means of relation between texts.

3 The Innovation: Industrial Research Laboratory

Where It Originated from

The second half of the nineteenth century saw great transformations within American society – and it was this environment that housed the corporations that first invested in research laboratories. These specific configurations of firms with their environment, which formed important factors in bringing about research laboratories in industry, need to be scrutinized. Several facets are of key importance: the emergence of the large, fully integrated, multi-divisional corporation, how these corporations were impacted by the legislative environment, and the expansion of higher education and professionalization in science.

The seeds for the birth of the giant corporation were laid in the dramatic expansion toward the west coast of the American continent, which necessitated both transportation and communications that could bridge an entire continent. Soon, railroads and the telegraph emerged as answers to these challenges (Chandler 1965; Israel 1992). While the Colonial Era saw enterprises hampered by transportation and communication problems, which kept them small in size, the post-Civil War Era saw the rise of large-scale organization fueled by societal and technological changes (Galambos 1975, 6ff), and it was this Second Industrial Revolution that in turn leads to the Gilded Age. As a result, corporations could now serve a rapidly expanding nationwide market. Concurrently with market growth, society became increasingly urbanized, which advanced the shift from an agrarian economy toward a manufacturing one, that was characterized by mass production (Chandler 1959, 2ff). This new setting imposed itself on the organization of businesses and led to increasing bureaucratization, specialization, the development of managerial hierarchies and the multi-unit enterprise, which eventually resulted in what is known today as the large, vertically integrated corporation that Chandler scrutinized in-depth (Chandler 1977; Galambos 1984, 472ff). Thus, most organizational characteristics associated with large-scale enterprise were in existence by the 1880s (Galambos 1975, 6). With the merger movement of the 1890s and the early years of the twentieth century (Lamoreaux 1985; Chandler 1977, 331ff), the markets saw another shakeup: Caused by mass-production, rapid growth since 1877, and the ensuing depression of 1893, a wave of consolidations followed to combat sinking prices. These newly consolidated firms had a strong grip on their respective industries, resulting in oligopolistic market structures. The merger movement shows how economic changes were not the only factor creating the specific scenario in which corporate R&D was born (cf. Lamoreaux et al. 2003). The wave of mergers from 1895 to 1904 can be seen as a result of the passing of the Sherman Antitrust Act in 1890, illustrating the need to include the legislative context of business.

In this context, two areas are of prime importance: antitrust and patents. Prior to the passing of the Sherman Act, corporations were colluding, fixing prices, and using cartels or other forms of combinations and market sharing agreements to come out ahead of the competition. After Sherman, with their previous business strategies being outlawed, firms merged horizontally. Combining into "single, legally defined enterprises" (Chandler 1977, 333) was the only possibility for avoiding prosecution and keeping control of markets. As Mowery (1995, 157) points out, US firms found themselves in a unique legal environment, which strongly influenced business strategies (see also Freyer 1995). The development of a system of intellectual property protection was the other area in the legal framework that can be seen as an important factor characterizing the environment of the first R&D laboratories. The US patent system initially only allowed the assignment of a patent to an individual, who subsequently could sell or license their patent rights to a corporation, a practice that came to be increasingly widespread with the end of the nineteenth century (Lamoreaux & Sokoloff 1999, 22; Noble 1974, 133; see also Fisk 2009, 75ff, for the gradual evolution of invention from individual toward corporate property). In addition, a strong system of patent protection leads to the creation of a market for technology with its own organizations, such as patent agencies and specialized trade journals tasked with the diffusion of new technological knowledge (Lamoreaux & Sokoloff 1999, 22ff). As Noble (1974, 133f) points out, while the earliest patent pooling agreement was in place by 1856, patents only became a key element in business strategies at the turn of the century.

Changes in legislation influenced not only the world of business but also another rapidly emerging and transforming sector of American society: higher education and science. With the passing of the Morrill Act in 1862, federal lands were given to states offering study programs in agriculture, home economics, and the mechanical arts, the latter usually interpreted as engineering (Rosenberg & Nelson 1996, 89; Reynolds 1991a, 21; Bruce 1987, 330; Cohen & Kisker 2010, 105ff; see also Geiger 1998). These new agricultural schools came to be known as "land-grant universities", further strengthened by the Hatch Act of 1887, in which federal money was given to establish agricultural experiment stations, which were usually tied to the land-grant colleges. This illustrates, on the one hand, an increase in the federal patronage of knowledge-production, as further exemplified by the 1863 creation of the National Academy of Sciences (Kevles 2013). On the other hand, it shows how knowledge-production was pragmatic and application-driven, as these newly minted universities were to apply their expertise to the needs of the community (Rosenberg 1976, 150ff). Furthermore, it served to strengthen the position of the emerging universities in the system of American higher education, as the land-grant colleges turned to universities for scientists.

Indeed, the fundamental transformation in American education from 1850 onward proved to be the ascent of the university to a position of dominance in regard to the production and diffusion of scientific and technical knowledge, made especially evident by the establishment of graduate programs and the subsequent increase in graduate- and PhD-level scientists (Bruce 1972, 88f; Kohler 1990). While prior to 1860 the "learned world" lacked distinct organizational features and most intellectual life took place outside of universities (Shils 1979, 21), the land grants soon shook up the landscape, and not just directly through funding the founding of universities, (agricultural) colleges, or technological schools. Their effects reached beyond that, with private funds further intensifying the "boost" (Cohen & Kisker 2010, 115; Bruce 1987, 330) and many universities being established in the years to follow, such as Johns Hopkins (1876), Clark University (1887), and the University of Chicago (1890). As a consequence, traditional, long-established liberal arts colleges such as Yale, Harvard, Dartmouth, and Princeton could not ignore the move toward science and technology fueled by the new agricultural colleges, in turn adding programs of their own in the natural sciences (Bartlett 1941, 485).

Simultaneously, engineering education aggressively expanded, and this academic variant ultimately triumphed over the apprenticeship-type system of training new engineers (Reynolds 1991a, 22). Over time, universities began combining teaching and research into a dual function. Originally their purpose was mainly pedagogical, with the teaching function differentiating the university from other learned organizations of the time (Hawkins 1979, 288). The formation of combined teaching and research is a highly complex topic, as its emergence depended on a variety of factors such as the return of graduate-level scientists from Germany, who experienced the German culture of conducting research at universities and stipulated changes toward a similar mission for their American counterparts (Dennis 1987, 493ff; cf. Kohler 1990), as well as the fight for the hegemony of ideals of pure science. At the end of the nineteenth century there stood the unquestioned dominion of the university as the producer of knowledge and the organizational home of the scientist, as universities had effectively become a synonym for science in the public eye (Shils 1979, 32). At the same time, science itself became popularized through popular magazines such as Scientific American and similar periodicals (Bensaude-Vincent 1997, 320ff).

Within the universities, science as well as engineering grew increasingly specialized, as new (sub-)disciplines led to the establishment of new study programs (cf. Galambos 1979, 269). While around 1850, the earth sciences – such as geology or mineralogy – dominated, the next decades saw enrollment and graduates in the natural sciences – for example, physics and chemistry – expand dramatically, eclipsing the earth sciences by the 1880s (Bruce 1972, 69). Moreover, networks of scholarly communications and the institutional structure of science grew in size and differentiation, the late nineteenth century seeing a high number of scientific societies founded and scientific journals publishing their first issues (Shils 1979, 40; Bates 1945, 28ff). Professional organizations were established not only in science but also in engineering and the American society as a whole, illustrating an increase of professional consciousness and fragmentation into specialized groups (Rae 1979, 252; Galambos 1975, 8; Galambos 2010; R. Bruce 1972, 91ff). Ultimately, the years from 1850 to 1900 saw the expansion of science as an institution, the advent of the science-based university, and the professionalization of the scientist and the engineer (Owens 1997). For the first time in American history, the university produced a surplus of graduates, not all of whom could be absorbed into the expanding system of higher education, thus leading to a large number of highly trained scientists that were available for work in industry.

To conclude this brief foray into the transformations of the Gilded Age society, one last question remains: What kind of relationship existed between academia and industry, if at all? Evidence for these interrelations is anecdotal and cursory at best, with many of the same names being echoed throughout the historical literature, such as Samuel L. Dana, who seems to have been employed as a company chemist in a textile mill as early as 1834 (Taylor 1976, 273), or the chemist C. B. Dudley, who in 1876 seems to have been the first chemist employed by the Pennsylvania Railroad (Taylor 1976, 273; Hounshell 1996, 17). Other firms appear to have employed chemists or other scientists, such as Carnegie & Company in the steel industry (Bartlett 1941, 492ff). It is generally assumed that these scientists were tasked with routine analyses, standardization, and similar occupations (Wise 1997, 220; Hounshell 1996, 18), whereas product development and improvement at this time seems to have been the province of the skilled mechanic, and the inventor, who boasted hands-on knowledge gained from years of work in the machine shops of the railroad and the telegraph. In this "Golden Age of Invention" (Carlson 1997) inventors such as Edison, Bell, and Tesla reigned supreme not only in the market for technological improvements and novelties that corporations turned to but also in the public mind (Hughes 1989; Hounshell 1984; Usselman 1992). The most famous of these, Edison, even gave his inventorial and entrepreneurial endeavors an organizational home in Menlo Park, established in 1876, where he united a diverse group of scientists, engineers, machinists, and other skilled workers to focus on the business of invention (Carlson 1997, 2007).

Another new organization emerged toward the turn of the century: the scientific consultancy. The most famous is without a doubt AD Little of Boston, established by the MIT chemists Arthur D. Little and Roger Griffin in 1886 (Kahn Jr. 1986; Rhees 1987, 34ff). They offered chemical analyses, standardization, and quality control, as well as product development to a certain degree. Painting a detailed picture of the time is a difficult undertaking since many historical accounts imply a distinct division between science and the industrial enterprise not only organizational or social but also in the tasks and practices performed in the laboratories or machine shops. Since the kind of knowledge used – whether craft knowledge honed for a long time, or scientific principles applied to industrial problems – can hardly be distinguished at this time with the accounts available, and furthermore any historical account being affected by the theoretical perspective and basic understandings of science, engineering, invention, and so on (Dennis 1987, 484ff), any conclusions drawn from the situation described need to be tentative. Relevant to the question of the industrial research laboratory is the following: Inventors were the primary means for corporations to attain new products and processes – inventors who boasted a considerable prestige in the public eye. Yet the separation between the university-trained scientist and the corporation was not as vast as sometimes claimed, with men of science finding their way into industry in a variety of positions on the one hand, and on the other organizations employing scientists and offering science-based inquiries emerging in the last two decades of the nineteenth century. Many of those university-trained scientists had done their graduate work in Germany and brought with them the knowledge, and sometimes direct experience, of organized research in the chemical industries. In fact, the industrial research laboratory existed in the German chemical industry as soon as the late 1870s, with many of the major firms such as Bayer, BASF, and Hoechst employing groups of scientists and, by 1882, the research laboratory having "emerged as a clearly distinguishable organizational concept" (Homburg 1992, 110).[1] It remains to be seen what role the news of such an organization that came with the German-trained scientists to the United States played.

In summary, American society transformed considerably from 1850 to 1900: The West was unlocked, markets and cities grew, and the large corporate organization emerged. The university became the primary institution of higher education, with the sciences and professions now growing and differentiating themselves. Scientists moved toward jobs in industry, taking their notions of the scientific method with them. These transformations put into place the environmental conditions that set the stage for the birth of the industrial research laboratory. In the next step, to understand how the laboratory spreads successfully in this environment, the process of diffusion shall be analyzed with the help of quantitative indicators.

How It Proliferated

The Problem with Mapping R&D Growth

How did the industrial research laboratory proliferate throughout American industry – an institution entirely new in 1900, but described as "firmly established" by contemporary writers in 1913 (Electrical World 1913, 877). Before diving into the available data accounting for the growth of industrial research laboratories, a few words of caution are necessary since both the "R" and the "D" in R&D prove to be slippery concepts. "Research" itself is not the ahistorical concept, it is sometimes made out to be, but a category whose meaning is highly dependent on the time and context surrounding it (Godin & Schauz 2016), especially when used to measure research activity in industry. As Godin explains, before the research was defined properly, when trying to measure its spread a general strategy to deal with the unclear boundaries of the concept was to supply a list of activities "designed solely to help respondents decide what to include in their responses to questionnaires" (Godin 2006, 648), offering categories such as basic and applied research, engineering, testing, or design. The impetus was to find out where research happened, not necessarily what kind in particular (ibid., 648f). Soon, the old dichotomy of pure and applied research was subject to the efforts of categorization and measurement, but early attempts were hamstrung often by a lack of data, the need to use proxies, or troubles with drawing clear delineations between pure and applied research (ibid., 649; Godin 2003). Eventually, a great number of taxonomies came into being, utilizing plenty of synonyms for research: applied science, basic science, pure research, fundamental research, etc. (Godin 2006, 650, table 1). Thus, any definition of research is strongly tied to underlying models of knowledge creation and the utility and application of science, such as the linear model or the ideology of pure science (Kline 1995; Lucier 2012; Godin & Schauz 2016).

Similarly, "development" is a concept that underwent changes in definition and meaning over the decades. In its early years, development was a subcategory of research, while still a distinctively industrial practice, only shifting to a separate category after 1945 (Godin & Lane 2011, 5). Still, research and development are commonly measured collectively, joined in the new abbreviation of "R&D", for two reasons: Firms did not have detailed accounting practices separating the two categories since they were thought of as interrelated, and a merger of these two practices increased the amounts of money devoted to research, with research expenditures having become a highly politicized topic (ibid., 6). Indeed, both concepts had become connected to questions of identity, with academic and industrial scientists, engineers, and other professions all vying for interpretive hegemony. Moreover, policy was informed by commonly shared assumptions of what R&D is and how research and development, or science and technology, relate to each other (Pielke 2012).

Further problems arise when shifting focus to the companies housing laboratories: As the history of the four research pioneers mentioned later will illustrate, many laboratories grew organically out of previous departmental arrangements for testing and product improvements, making it hard for companies responding to official inquiries charting R&D growth to pinpoint when their research ventures started. With corporate laboratories becoming somewhat of a "must-have" for big companies following World War I (WWI), it is reasonable to assume that some simply started to refer to their machine shops and testing laboratories as proper research laboratories, regardless of what was happening inside.

Luckily, the symbolic and political struggles between different understandings of research and development are not of special concern at this point since the goal here is simply to get an idea of how the industrial research laboratory spread and came to be accepted as an organizational entity, and not what special brand of research was possibly conducted. Of course, the fashion of research poses a challenge for assessing laboratory growth, but in itself can serve as a result: Corporate R&D had become normalized and expected, an outcome informed not only by an increase in corporate establishments that did research but also by cultural and discursive shifts in the meanings of science and the scientist. In the end, all quantitative measures discussed in the following need to be considered with reservation with regard to their contextual sensitivity, their method of data gathering, and their underlying definition of research laboratory.

Sketching the Growth of Industrial Research Laboratories

Few data points exist for the early years of organized R&D. The major source that quantitative studies draw on when outlining how industry transformed are surveys carried out by the National Research Council (NRC). These surveys were conducted for the years 1920, 1921, 1927, 1933, 1938, 1940, and 1946 (NRC 1920/21,[2] 1927, 1931, 1933, 1938,

1940a, 1946), based on questionnaires sent to directors of industrial research laboratories. To compile lists of laboratories, scientific societies and their respective journals were consulted, as well as trade associations such as the National Association of Manufacturers (NAM). Furthermore, trade journals and industry catalogs were examined as well (NRC 1921, 2). In the questionnaires, the respondent was asked to provide data on the founding date of the laboratory, the parent firm of the laboratory, the location of the laboratory within the firm, and the number of employees (both scientific and non-scientific) and their respective disciplines, as well as information on what kind of research work was undertaken and whether the scientists worked full- or part time, and with what kind of equipment (NRC 1920, 47; Mowery 1981, 48). The surveys mainly covered laboratories of manufacturing companies, yet some independent contracting laboratories were included as well.

The NRC simply let the companies themselves classify what they understood as research (Perazich & Field 1940, 2; see also Godin 2006, 647f), adding the following caveat: "Most of them devote but a portion of their effort to research, and a number are probably not research laboratories, under a strict definition of that word" (NRC 1920, 45). This is justified by the unclear boundaries of science in industrial employ:

> Research is sometimes differentiated into 'scientific' and 'industrial'. Scientific research comprises investigation directed toward the discovery of new truths for the sake of increasing human knowledge. Industrial research is the endeavor to learn how to apply scientific facts to the service of mankind. Many laboratories are engaged in both industrial research and industrial development. These two classes of investigation commonly merge so that no sharp boundary can be traced between them .... Furthermore, in practice it frequently is difficult to keep clear the distinction between scientific and industrial research.
>
> (NRC 1920, 45f)

The first-hand information supplied by the companies was not checked for accuracy: "No investigation has been made to ascertain the character of any laboratory listed, nor the quality of the work done" (NRC 1920, 46). The various issues of trying to define "research", as outlined above, add a large amount of fuzziness to the data, especially in regard to the question of whether the surveys really measure the diffusion of industrial research laboratories or just the spread of the name for any kind of department that did any kind of research, testing, engineering, or development work. As Cooper (1940, 173) notes, "the individual returns reflect the diversity of research activity throughout the country, and illustrate, among other things, the looseness of definition of the term 'research'".

Apart from the definition of research, there are further problems with the NRC data. The degree of coverage varies over the years, especially for the first survey conducted in 1921, but is considered complete for the surveys since 1927 (Perazich & Field 1940, 2). Smaller, harder-to-find laboratories are possibly omitted (ibid.), skewing the data toward large laboratories. Due to the nature of the questionnaire, differences in reporting of data are known, particularly concerning the different types of staff, such as technical, non-technical, scientific, etc. Lastly, the earliest surveys do not give the foundation dates of the laboratories. These dates were only reported in the surveys from 1940 onward (Mowery 1995, 153n8), though Cooper (1940, 273), using the NRC survey data, has a complete dataset of laboratory creation per year from 1890 to 1940, although it is not clear how he compiled the numbers for the early years. It is obvious, then, that the NRC survey data have to be assessed carefully and in no way can be used to paint an accurate picture of diffusion. Still, they can surely be useful in charting scientific employment in industry, as well as the spread of the notion of "research".

Another layer of data may be called "qualitative" data on laboratory foundation. On the one hand, there are reports compiled of industrial laboratories, similar to the idea underlying the NRC surveys yet far smaller in scope, that describe industrial research as exemplary and in-depth rather than giving an exhaustive picture of the entire United States (Fleming 1917; Angell 1919; NRC 1940b). On the other hand, much of the academic literature on the origins of industrial research mentions the diffusion of laboratories in some way or another, as well as early adopters of in-house research, apart from the research pioneers. Usually, key companies are named, but the beginnings of their research efforts vary widely from publication to publication, as well as the numbers reported (Hounshell 1996, 21, 36; Chandler 1977, 375; Weart 1979, 306; Thackray et al. 1985, 114; Sturchio 1981, 87f; Carlson 1997, 217). Other publications compile writing on these laboratories, for example, West (1930), who compiled "a reading list of selected articles from the technical press" for the Division of Engineering and Industrial Research of the NRC, with the goal of offering insights into how laboratory research should be organized and what results can be expected (ibid., 5). The fact that large bodies of literature were written on industrial research certainly speaks for the organizational entity gaining visibility and acceptance. What then can be learned from the NRC survey data and respective quantitative analysis undertaken? Since the discourse analysis will focus on the discipline of chemistry, special attention shall be given here to the chemical industry.

Laboratories and Their Parent Companies

As can be seen in Figures 3.1 and 3.2, the different publications report strongly varying numbers, leaving the exact number of laboratories founded between 1900 and 1930 up in the air. Yet clear trends can be recognized. The number of industrial research laboratories grew steadily in the five decades between 1890 and 1940, with the strongest growth occurring between 1919 and 1928, right after WWI and roughly to the beginning of the Great Depression, which ended the postwar surge. The chemical industry accounts for one-fourth of all laboratory foundations between 1899 and 1946, and the very "research-intensive" industries of chemicals, glass, rubber, and petroleum founded about 40% of all laboratories in the period (Mowery & Rosenberg 1989, 61; Mowery 1981, 53f). Contrasting other manufacturing sectors, the chemical industry proved to be a leader in research across all time periods, whereas other sectors – such as instruments, transportation equipment, and machinery – got a late start (Mowery & Rosenberg 1989, 61).

Generally, large amounts of capital are seen as a prerequisite for the establishment of laboratories, but as Mowery (1983b, 965) shows, outside of chemicals, large firms were not disproportionately intensive employers of scientific personnel. Instead, he found that large, non-chemical firms employed significantly fewer research professionals than small firms, relative to their respective sizes. In contrast, in the chemical

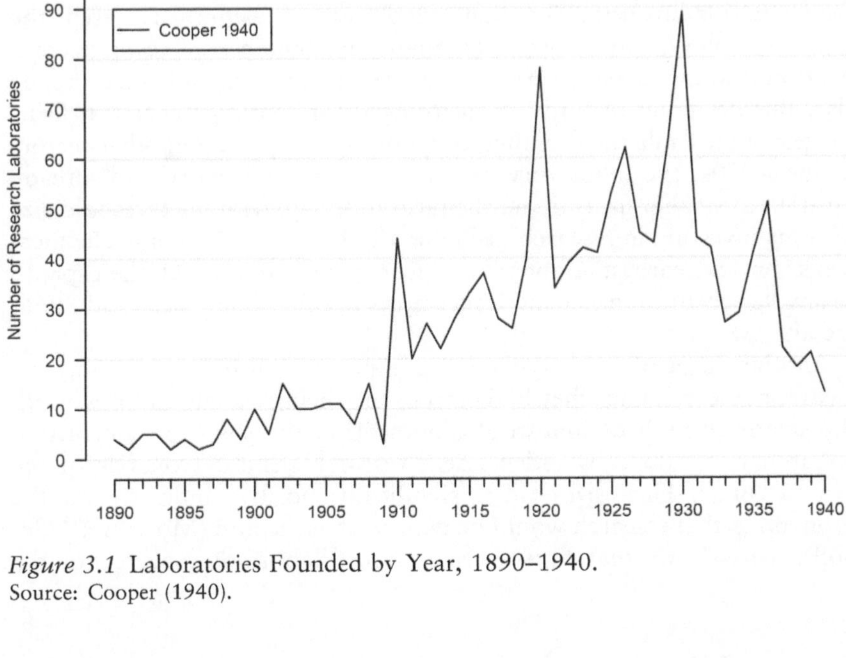

Figure 3.1 Laboratories Founded by Year, 1890–1940.
Source: Cooper (1940).

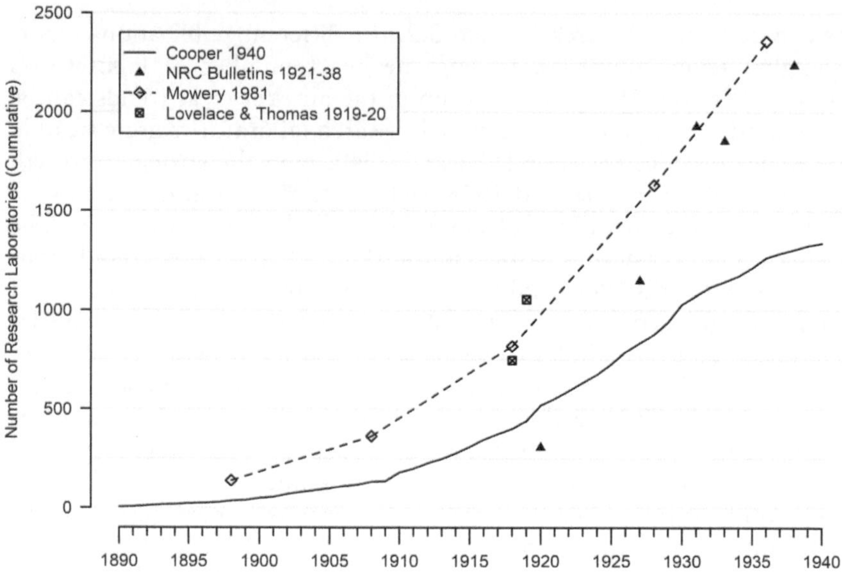

Figure 3.2 Cumulative Growth of Industrial Research Laboratories, 1890–1940.
Sources: Cooper (1940), NRC (1920, 1921, 1927, 1931, 1933, 1938), Mowery (1981), Lovelace and Thomas (1919, 1920).

industry, large firms had the biggest laboratories and proved to be the most research-intensive from 1921 to 1946, likely due to the nature of the chemicals business. The high, fixed costs of chemical research, the large portfolio of diversified companies, and the absence of price competition lead the largest firms to invest heavily in research before anyone else. But declining research intensity even for the large chemical firms suggests that while the founding of laboratories sped up in other sectors in the 1920s, the laboratory as an organizational entity had already diffused to smaller firms in the chemical industry, with the threshold size of firms investing in research declining (ibid., 964, 967). Large chemical firms were the earliest adopters of industrial research, which then spread to smaller chemical firms and larger firms in other industries (ibid., 966; see also Mowery 1981, 177ff).

Another facet of note is whether companies still bought outside services for research once they had established their own laboratories. With the steady growth of industrial laboratories, the share of contractual research being done by independent research institutes outside of the firm declined. Firm size had a significant, positive influence on the probability that research would be done within the firm (Mowery 1983a, 369). This shows that firms with in-house laboratories were not the primary clients for contract research organizations (ibid., 363), yet it seems like sometimes both services were used, suggesting that in-house research complemented rather than supplemented contract work (ibid., 363ff), with many contract laboratories doing standardization, testing, and quality assurance, whereas in-house laboratories were concerned with development.

Staff, Scientific Disciplines, and Expenditures

Between 1921 and 1946, the number of scientific personnel increased sharply, with the greatest increase happening between the years 1921 and 1927, concurrent with the boom in laboratory foundations after WWI. Only the Great Depression put a temporary damper on personnel growth, which regained traction after 1933. Factors that influence the increase in staff are, first, that the growth in staff is due to the increase in personnel employed at the existing laboratories (cf. Cooper 1940, 175; Mowery 1981, 77). Second, new laboratories were created leading to staff growth as well. Third, some of the staff growth can certainly be attributed to NRC survey coverage in later years. Fourth, recent surveys included additional classifications for research personnel, further inflating the staff numbers. Still, even with some of the growth accounted for as data artifacts, a clear trend of increasing research employment can be seen. Furthermore, the employment of part-time researchers declined.

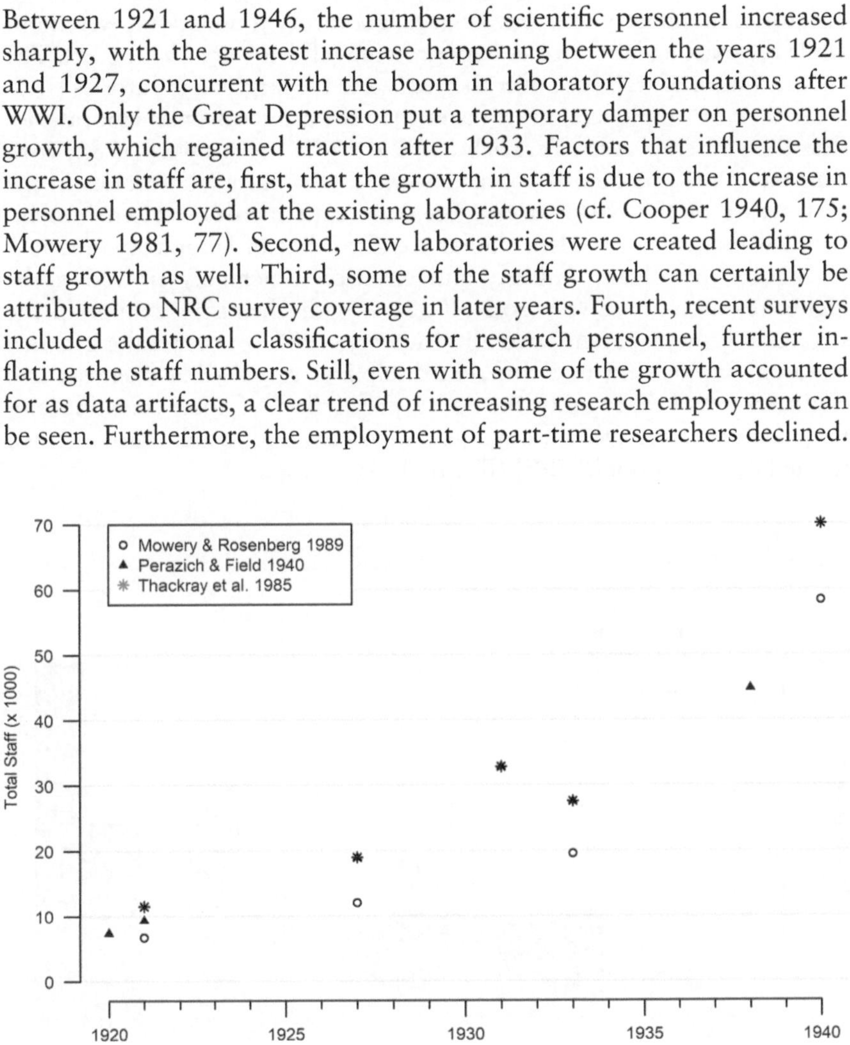

Figure 3.3 Total Staff Growth.
Sources: Mowery and Rosenberg (1989), Perazich and Field (1940), Thackray et al. (1985, 347, table 5.5).

Perazich and Field hypothesize here that research turned from an expensive luxury to a necessity, making fully employed science workers out of researchers who had previously devoted only parts of their time, for example, during universities' summer break, to research (Perazich & Field 1940, 5f).

While the growth rate of newly founded laboratories declined over time (Figure 3.1), research personnel kept increasing (Figure 3.3). This is most likely due to only a finite number of companies existing that would organize their own laboratories. Thus, once a certain "saturation" of laboratories in industry was reached, the laboratory foundation petered out while the existing laboratories kept growing (cf. Mowery 1981, 63ff). The concentration of staff rose over the years and the largest laboratories (with more than 50 employees) grew faster than the group of small laboratories (less than 11 employees; Perazich & Field 1940, 66; Mowery 1981, 76f). This effect may be due to poorer coverage of smaller laboratories in the early years of the NRC survey. Breaking staff increases down into individual industries reveals a similar trend to that of laboratory creation: Chemicals, petroleum, rubber, and electrical machinery proved to be highly research-intensive across all years, with chemistry exhibiting the strongest growth pattern overall (Mowery 1981, 61f). Breaking down research employment into different disciplines reveals that chemists and engineers retained roughly the same proportion of total scientific employment, whereas other groups (e.g. physicists and metallurgists) gained in the later years (ibid., 79ff) (Figure 3.4).

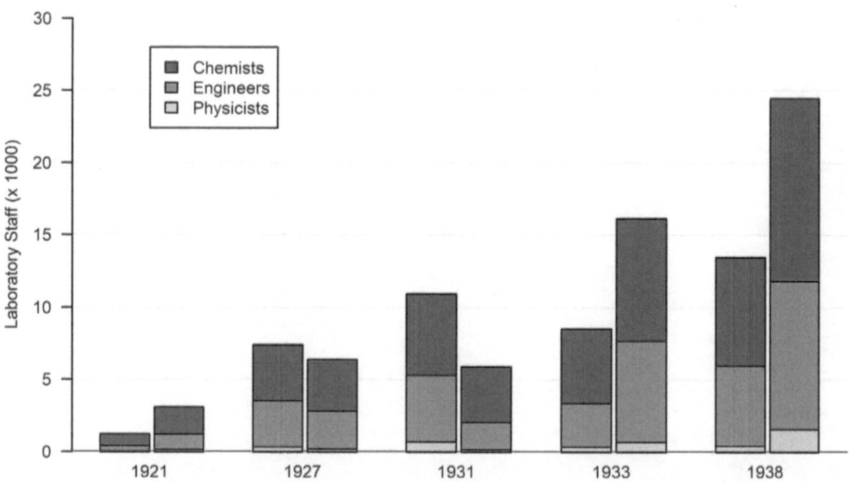

Figure 3.4 Staff by Scientific Disciplines Employed in Laboratories, 1921–1938.
Sources: Left bars: Perazich and Field (1940, 78, table A-19), right bars: Thackray et al. (1985, 347, table 5.5).

The Decline of the Independent Inventor

Before taking a detailed look at the pioneering corporate laboratories, another indicator that is considered to be highly related to industrial R&D shall be briefly scrutinized: the independent inventor. It is generally assumed that the era of heroic invention ended with corporate laboratories (Hughes 1989, 138f; Noble 1974, 151ff), as the university-level scientist replaced the self-trained inventor. Hintz uses statistics on the number of patents issued to independent inventors versus those issued to corporations as a measure to gauge the fate of independent invention. As his time series from 1901 to 2000 shows (Hintz 2011, 734, figure 1), the demise of the independents was not as fast as generally assumed, with the number of patents taken out by corporations only overtaking individual patentees in 1933. Until the 1950s, independent inventors still accounted for a major share of the patents assigned, and only then did corporate dominance of patenting begin. Thus, the independent inventor was not abruptly displaced by the corporate laboratory, rather a long transitional period began at the turn of the century (cf. Wise 1992). Nicholas corroborates these findings, as his dataset shows a steady decline of independent patentees since the 1880s, yet even in the first three decades of the twentieth century – when corporate R&D is considered to have "taken off" – a considerable share of patents was assigned to independents (Nicholas 2010, 58, figure 1). Interestingly, focusing on chemical and electrical patents reveals those corporations already overtaking independents by the late 1920s (ibid., 77, figure 3). Nicholas' data lead to the conclusion that firms utilized independent inventors as complements to their own R&D (ibid., 78), and in a similar way to contractual research, as was discussed earlier.

Lamoreaux and Sokoloff take a more nuanced approach to the question, differentiating to whom patents were issued between 1870 and 1911: an individual, a partner, a corporation with a similar name, or a corporation that was known to have a laboratory, among others. The latter two categories allow the distinction between what is often called the inventor-entrepreneur, who chose to go into business with their own inventions as a principal rather than an employee, and the inventor who worked for a corporation with a research laboratory (Lamoreaux & Sokoloff 2009, 63f). Constructing an elaborate dataset, they found that over time the association between specialized inventors and company assignments strengthened (ibid., 66), indicating the growing importance of R&D laboratories. But not all results match this pattern: Over time, an even greater proportion of patents was assigned to companies bearing the inventor's name, illustrating that work in a company laboratory was not the only option for inventors at the turn of the century. Yet the contractual mobility of inventors declined, which means that inventors tended to stay with one company throughout their career. This suggests that inventors deemed it necessary to form long-term attachments for a variety of reasons.

The end of the era of heroic invention proved to be a slow, drawn-out process, rather than a revolution brought about by big business. The results from the studies cited above show that there was more to their decline than simply a new mode of corporate innovation immediately replacing the inventor with the scientist. A cultural dimension was strongly involved in this process, pitching the genius inventor against the scientist, each with their respective promises regarding their benefits for the company, with the scientist's reputation slowly building. Over time, also in historical literature, the scientists seem to have played the winning game (Hintz 2011). It is precisely this dimension of identity ascribed to the scientist, or more specifically the corporate scientist, that shall be investigated through methods of discourse analysis in order to further elucidate how and why the patterns we perceive in the quantitative data were produced.

The Research Pioneers

Of the plethora of industrial research laboratories established by 1930, four came to be regarded as pioneers. These were the laboratories of General Electric (GE), Du Pont, AT&T, and Eastman Kodak. These laboratories, established within the first 12 years of the new century, became highly visible and successful due to scientific breakthroughs, cultivating archetypal actors of heroic scientists and gifted research administrators in their wake. They also became the best-studied literature, for reasons that will be discussed later.[3] Although all these laboratories were established around the same time, a variety of reasons exist and differing strategies were chosen to create those in-house R&D efforts, sometimes from scratch, and sometimes from the gradual growth of existing arrangements for product innovation (Kline & Lassman 2005, 605). Furthermore, some authors argue that, in a way, each of these firms re-invented managerial strategies as to how to lead and structure a corporate R&D laboratory (Dennis 1987, 488). This is precisely why a closer look at those highly publicized pioneering efforts is needed – to see what can be learned from them before tackling the cultural dimension that is lacking in most explanatory models, especially since many important concepts, topics, and issues of the day will be introduced with these examples. For each company, we are interested not only in the factors involved in the creation of the laboratories but also in general explanations for their emergence found in historical scholarship on those topics, so as to assess the level of current scholarship and provide opportunities for possible blind spots to emerge, as well as to underline the importance of the perspective and methodical tools this book adds.

1900: GE

In the literature on the genesis of industrial R&D laboratories, GE is generally considered to be the trailblazer for all the companies that followed suit in the first two decades of the twentieth century. In order to fully comprehend the circumstances that led to the establishment of this laboratory, two important building blocks are of relevance: the situation and currents of the electrical industry in which the GE Company was founded, as well as GE's organizational structure that served as the cradle of the laboratory.

The Electrical Industry in the Late Nineteenth Century and the Birth of GE

The electrical industry went through its formative period from 1880 onward (Bright 1949, 70ff). It was characterized by rapid technical change due to many competing lighting technologies and lamp designs, as well as continuous improvements to existing technologies and manufacturing techniques. Next to competing lighting models, a "war of the currents" (Reich 1985, 46; Carlson 1991, 283ff) was fought for the standardized current of electric power supply (AC versus DC). Patent struggles among a multitude of competing manufacturers put a damper on technological development – thus, with established design standards, production and marketing played a vital role. The companies themselves utilized buy-outs, mergers, legal suits, and collusion as means to suppress competition (Reich 1992, 307; cf. Carlson 1991, 278ff).

The inventor-entrepreneur played an important role in this industry, in a time sometimes referred to as the "Golden Age of Heroic Invention" (Carlson 1997, 206ff; Hounshell & Smith 1988, 3; cf. Usselman 1992; Wise 1992), which is most commonly associated with the figure of Thomas Edison, who was a key player in the formation of GE and the founder of Menlo Park, the "invention factory", in 1876 (Hounshell & Smith 1988, 3).[4] In 1889, the Edison General Electric (EGE) Company was founded by reorganizing Edison's many distinct companies into one single corporation. Two other important actors in the electrical industry emerged at this time: Elihu Thomson and George Westinghouse. The latter entered the field in 1896 with an alternating-current system (Reich 1985, 46; on Westinghouse, cf. Usselman 1992; Kline & Lassman 2005), becoming the main competitor of GE in its early years. The former, Elihu Thomson, became one of the founding members of GE.

Elihu Thomson, a chemistry teacher from Philadelphia, was one of the most prolific inventors in American history (Carlson 1991, 1). Previously a professor of chemistry at a college in Philadelphia, a radical change of profession to invention (ibid., 53) saw him, together with Edwin J. Houston, develop a system for lighting (Carlson 1997, 208). After these first failed business experiences, the Thomson-Houston Electric Company (TH) was founded in 1882, by "a group of shoe manufacturers in Lynn, Massachusetts, to exploit the inventions of Elihu Thomson and Edwin J. Houston" (Carlson 1995, 61). Led by Charles A. Coffin, an expert at marketing, Thomson could concentrate on inventing, leading to the development of new products, such as AC and DC incandescent lighting systems, also motors and streetcars that subsequently allowed the company to expand and compete with EGE and Westinghouse Electric (Carlson 1997, 209).

EGE's and TH's competitive activities ultimately led to their merger in 1892, forming GE with its head office in Schenectady, NY. The merger was seen as a lucrative move due to both companies' strong patent positions and their complementary manufacturing lines, as well as EGE's capital and TH's managerial capacities (Birr 1957, 30; Carlson 1991, 299; Carlson 1997, 209). Edison soon pulled back from business operations (Reich 1985, 48), and the managerial group of TH then comprised almost the entire management of GE, with Coffin becoming the company's first president (Carlson 1991, 296f). Coffin took the company's organizational "building blocks" of EGE and TH and merged both into a unified structure, its hierarchy resembling that of a large railroading company (ibid.).

The newly formed company soon became known as the "electric trust", due to its strong position in the market (ibid., 50) and its anti-competitive activities (Reich 1992, 306). By the mid-1890s, GE and Westinghouse controlled approximately 75% of the electrical industry (Reich 1992, 308; cf. Carlson 1991, 274, figure 6.1 on the evolution of both companies). Their main tactics against competition were pricing and patent infringement suits that were ultimately put aside in a licensing agreement between both companies in 1896, further strengthening GE's market position (Carlson 1991, 278). In 1897, GE organized a cartel, the Incandescent Lamp Manufacturers Association, which allocated 50% of the market to GE and allowed for easy control and little need for improvements in the company's technological base (Reich 1992, 308).

Shifting our perspective from the market environment of GE toward its organizational structure, it is important to understand where product and process improvements took place. The company engineers were organized in different groups, and at Schenectady, a group called the "Works Laboratory" did materials testing. In 1893, GE bought the company belonging to Charles P. Steinmetz, an "engineering genius" (Reich 1985, 59) who came to GE's attention due to his papers being presented before the American Institute of Electrical Engineers and his highly innovative work for a New York electrical manufacturer. Steinmetz joined the Calculating Department where he brought "the theories of electrophysics to bear on problems in engineering design" (ibid). The Standardizing Laboratory was established in 1896, and its role expanded in the following years toward the design of new instruments and the creation of new manufacturing standards. As can be seen, many different groups worked on routine tasks as well as continuous improvements – an organizational setting befitting the corporate strategy for innovation of the time. Patents and inventions were bought from outside inventors, and their services were only used temporarily without any fixed or continuous arrangement (Reich 1985, 61).

But large transformations were inbound, heralded both in- and outside the company. Thomson, working as an inventor within GE, and being an important voice as one of its founders, grew increasingly frustrated with the way that product development was handled and started to advocate for the hiring of specialists, whereas proposing the establishment of a permanent department, isolated from day-to-day operations, in order to coordinate resources (Carlson 1997, 211). On the outside, GE's secure grip on the market was about to slip away: the expiration of all of GE's electrical patents by the end of the century loomed large. European inventions also threatened GE's market position (Reich 1985, 61), as well as continuing competition with Westinghouse and the ubiquitous danger of new, more efficient lighting devices coming to market (Carlson 1997, 212; Wise 1987, 396).

Pushing for Change: Establishing the Laboratory

The story of the laboratory's establishment is a story of actors in the organization attempting to advance their interests. Chief engineer Steinmetz was well aware of technological developments in the field, and in 1897 proposed organizing a chemical laboratory to investigate new, competing devices, possibly informed by his experiences studying in Germany and Switzerland.[5] This request was ignored by GE officials (Carlson 1997, 212), a similar plea in 1898 did not yield the desired result either (Reich 1985, 65). After his first two requests fell to deaf ears, he enlisted support for his plea: Albert G. Davis, GE's patent attorney, and Edwin W. Rice, the vice-president in charge of manufacturing and engineering. Both Steinmetz and Davis wrote urging letters to Rice, who used these to convince GE president Coffin and the board of directors of the need for the new department (Reich 1985, 66; Carlson 1991, 336f). Davis argued from an economic angle: "'If someone gets ahead of us in this development [the mercury-vapor lamp] we will have to spend large sums in buying patents ... whereas if we do the work ourselves this necessity will be avoided'" (Davis, quoted in Reich 1985, 66). In his letter, Steinmetz described a detailed research plan for the laboratory, and suggested the laboratory be separated from manufacturing and production problems and directed by a "practical chemist" (ibid.). Their advance gained even more traction, once Thomson threw his weight behind the proposal, voicing his support to Steinmetz and Rice.

In his letter to Rice, he stressed that the company should research the application of new principles and even the discovery of those principles as well (ibid., 67), devising a broader mission for the laboratory than Steinmetz.

The coalition of interest managed to convince the executives, and the Board of Directors approved the plan under the condition of finding the right man: "'With such a man this may be a fine thing, without him, it may simply be a machine for spending money'" (GE Board of Directors, quoted in Reich 1985, 67). Rice picked Willis R. Whitney, a professor at MIT who had done his doctoral work in Leipzig, Germany under Wilhelm Ostwald, one of the fathers of physical chemistry (Wise 1983, 10ff), and who had some experience with scientific work outside of academia: He had done some scientific consulting in the 1890s (Wise 1985, 58). After being interviewed by Thomson, a compromise was worked out: Whitney would remain at MIT due to his strong devotion to teaching, but spend two days a week at GE during the school term, earning the salary equivalent of a full professor (Reich 1985, 69). By 1901, Whitney had requested a leave from MIT to become the full-time director of the laboratory. Ultimately, he would become the archetypal research director (Wise 1985; Reich 1985, 97ff; cf. Broderick 1945). The laboratory was erected at GE's Schenectady plant, virtually "from scratch" – as Kline & Lassman (2005, 605) note, with neither Edison's laboratory nor Steinmetz's engineering department serving as templates for the research laboratory. In its annual report for 1901, GE officially heralded the new laboratory, with Rice stressing the laboratory's exclusive devotion to original research: "'It is hoped by this means that many profitable fields may be discovered'" (GE Annual Report 1901, quoted in Reich 1985, 71).

The First Decades of the GE Research Laboratory

The newly created laboratory grew quickly: its staff numbered about 50 employees in 1905, about 100 in 1910, steadily climbing to a peak in 1929 of about 520 employees, before the Great Depression forced staff cutbacks, down to about 280 in 1932 (Wise 1985, 246, figure 2). Laboratory expenditures described a similar trajectory after a slow start, doubling roughly every four years and peaking in 1929 at about $2.6 million (ibid., 246, figure 1).

In its early years, the research laboratory struggled to produce major breakthroughs, yet fulfilled several corporate needs by working on special development requests made by the manufacturing and engineering departments (Reich 1985, 70f); by 1903 Whitney and his group had carved out a "reasonably well-defined – if not necessarily secure – niche in the GE corporate structure" (ibid., 73). Differing approaches to research work led to a control struggle between chief engineer Steinmetz, in charge of the oversight committee, and research director Whitney. In 1905, William Coolidge was hired, who had graduated with a B.S. in electrical engineering from MIT, and a PhD in physics from the University of Leipzig, Germany. A high salary offer and opportunities to spend one-third of his working hours on his own research swayed him to start at the GE laboratory (Reich 1987, 345ff; Wise 1983). GE was fully engaged in a race against European inventors and Westinghouse to develop better and more durable lamps. Results were requested from the laboratory, which reported its first success when Coolidge managed to show how tungsten could be made ductile, but this achievement only came after GE was forced to acquire German patents for a tungsten-filament lamp in 1906 (Carlson 1997, 213).

These troubles, as well as a market panic in 1907, led president Coffin and vice-president Rice to cut the laboratory budget by 41%, thereby terminating one-third of the professional staff of 44, which caused Whitney to suffer a nervous breakdown and Coolidge to take over the reins during the former's recovery (Reich 1985, 79; cf. Wise 1985, 245, figures 1 and 2). Still, the lab's development of the ductile tungsten filament could be regarded as the fulfillment of the "basic mission for which it had been established" (Reich 1985, 82). With the newly gained advantages that the laboratory brought, GE continued its control of competition during the early twentieth century, even though the other major competitor, Westinghouse, had ceased to pose a serious threat to GE since its bankruptcy reorganization that was brought about by the Panic of 1907 (ibid., 313).

Of the many scientists and engineers working in the laboratory over the years, one other figure stands out: Irving Langmuir, who was hired in 1909. Langmuir graduated with a B.S. in metallurgical engineering from Columbia University, doing his graduate work in physical chemistry at the University of Göttingen, Germany. Langmuir's work led to breakthroughs in lamp efficiency, later spearheading the company's forays into X-ray and radio, and eventually culminated in a Nobel Prize in chemistry for research on physical processes at the surface of incandescent filaments for Langmuir (Reich 1987, 347f; cf. Wise 1983; Reich 1983). To lure brilliant young researchers such as Langmuir into industry, Whitney offered significantly higher wages and better equipment than academic institutions. Publishing results was allowed but only after patent applications, and joining scientific and engineering societies was encouraged, with both Whitney and Langmuir later serving as presidents of the American Chemical Society (ACS; Reich 1985, 108ff; Birr 1957, 80; Wise 1997).

Examining GE's organizational structure during those years, a number of other laboratories can be found: the Standardizing Laboratory (later merged with Steinmetz's group to become the General Engineering Laboratory), the Schenectady Works Testing Laboratory, the Illuminating Engineering Laboratory, the Insulations Laboratory, and the Consulting Engineering Laboratory (run by Steinmetz from 1909 to 1919), whereas others were established as needed, for example, the Radio Engineering Department that started work in 1918 (Reich 1985, 104). These additional departments relieved the research laboratory of most of the routine investigations and development work.

From roughly from 1911 onward, the research laboratory's work diversified into many different project areas, but the staff remained "on call" to solve problems occurring in engineering or production departments (ibid., 91; cf. Carlson 1997, 216; Birr 1957, 51). Defensive research aided GE in covering the market and keeping other companies from catching up to its technological base (Reich 1985, 86).

After the United States entered WWI in 1917, the research laboratory was engaged with wartime research, such as work on submarine detection and other projects (ibid., 93; Birr 1957, 63ff). Whitney served on the Naval Consulting Board as a representative of the ACS, not shying away from advertising the efficacy of industrial research, and in the process brought the laboratory to the attention of many other business executives and managers, but also the public (Reich 1985, 94). Science's newfound popularity was soon exploited by GE as an advertising opportunity: the research laboratory became known as the "House of Magic", permanent science exhibits were set up and public tours were organized (ibid.; Wise 1985, 214f; Hawkins 1950, 78; see also Nye 1985).

1902: Du Pont

Chronologically, the next company to establish a laboratory was Du Pont. To understand how Du Pont got to have not one but two laboratories, a detailed discussion of the company's leadership and structure is necessary, which sets the stage for the laboratories' successes and later shift to "fundamental research" in the late 1920s.

The Field of Chemistry and Du Pont's Organizational Antecedents

Founded in 1801, E.I. Du Pont de Nemours & Co. had a long prehistory as one of the leaders of the American gunpowder industry (Wilkinson 1984; Chandler 2005, 42). The company's transformation started in 1902 with the death of Eugene du Pont, the company's president. Prior to 1902, Du Pont had existed as a family partnership essentially made up of many different powder plants that received little oversight or administration, and hardly any attention was paid to "costs, to improving processes, or to developing more systematic purchasing and marketing techniques" (Chandler 1990a, 54). Improvements seem to have been made through experimentation and hands-on-work at the various plants, and there are no records that indicate any organized research prior to the company's reorganization. In the resulting struggle for control within the family after Eugene du Pont's death, Alfred I. du Pont emerged, who would lead the company with his cousins Pierre S. du Pont and T. Coleman du Pont. With all three having studied chemistry or engineering at MIT, they had gathered practical, managerial experience in various plants and powder mills. The three du Ponts pursued a strategy of consolidation and centralization for the various companies controlled by Du Pont, at the same time creating an administrative structure to oversee the newly incorporated E.I. du Pont de Nemours & Company in 1902. In the reorganization, T. Coleman became company president, Pierre S. vice-president and treasurer, and Alfred I. assumed the duties of general manager.

The three cousins were guided by their mentor, Arthur J. Moxham, who had gathered management experience in the steel industry. Moxham convinced the cousins to stop the practice of buying a majority interest in other explosives companies and running them as separate entities, and instead to purchase companies outright and build a multidivisional structure under the guidance of an Executive Committee (Chandler & Salsbury 1971, 125ff). Thus, the Executive Committee was established in 1903, initially comprised of the president and the department heads. The Committee would coordinate the various departments, set policy, and approve expenditures. Alongside the cousins, Francis I. du Pont, Moxham, a sales expert named J. Armory Haskell, and a veteran of the explosives business named Hamilton M. Barksdale formed the Committee (cf. Hounshell & Smith 1988, 19, figure I.I). Furthermore, Moxham was chosen to head the new Development Department, tasked with improving products and processes – an organizational element having "no predecessor in the explosives industry", as Chandler notes (Chandler 1990a, 57). It was then decided that the companies routine concerns should be dealt with by an Operative Committee, whereas the Executive Committee would focus on long-term planning, steering clear of departmental day-to-day operations. The organizational structure created in this way remained mainly the same until the end of WWI, save for the creation of some auxiliary departments (Chandler 1990a, 62, chart 2).

Establishing the Laboratories

Today two names have become synonymous with Du Pont's early R&D operation: the Eastern Laboratory, and the Experimental Station. In 1902, Repauno, a Du Pont subsidiary, founded the Eastern Laboratory – named after its parent holding company, Eastern Dynamite. Repauno's managers "had realized that 'chemical work' was necessary to maintain and improve product yields and quality" (Hounshell & Smith 1988, 19f). Oscar R. Jackson, a Harvard and German-trained chemist, became the new plant superintendent, subsequently hiring academically trained chemists to do some experimental work. The research happened in a "works laboratory", which was not separated from manufacturing. Its positive results convinced Repauno's management – among them Barksdale and Haskell, who went on to form part of the Executive Committee – to establish the Eastern Laboratory. Both Barksdale and Haskell were of the conviction that chemical work was worth the effort to keep a leading position in the explosives industry (ibid., 21).

Thus, the Eastern Laboratory was born, and its "broad mandate" (ibid., 19) was made explicit in its mission: improving products and processes, discovering new explosives, keeping in touch with developments in the industry, as well as training chemists and technical assistants (ibid.). It was divorced from the day-to-day manufacturing operations of Repauno but was expected to do inquiries for all the plants under Eastern Dynamite's roof. To direct the laboratory operations, Charles L. Reese was hired, a University of Virginia graduate, who had done his PhD work in Göttingen and Heidelberg, Germany. It was Reese who established the Eastern Laboratory's research policy, instituting a report system and hiring the early laboratory staffers (cf. ibid., table 1.1). Under Reese's direction, this group of scientists could record early successes, demonstrating the usefulness of the laboratory (ibid., 26).

Du Pont's other laboratory, the Experimental Station, originated from an entirely different point in the organizational structure: the Executive Committee. It was Moxham, in charge of the Development Department, who convinced his colleagues to create a General Experimental Laboratory under his care in 1903. To do this, he first had to establish a broad base of support within the company (ibid., 29), forming an "Experimental Committee" that was staffed by himself, Alfred I. and Francis du Pont, as well as Barksdale. In this Committee, he proposed a clearly outlined research project with obvious benefits to Du Pont's business, also referencing successful laboratories in Germany and England, which ultimately garnered the necessary support from his colleagues. At first conceived to do "original research" (ibid., 30), it was Barksdale – seeing this new laboratory as a direct threat to Eastern – who urged the Executive Committee to restrict the new laboratory's domain to the screening of outside inventions (ibid.). It would take outside intervention to refocus the new laboratory. Late in July of 1903, Francis I. du Pont informed the Executive Committee about the Navy establishing a smokeless powder plant and providing R&D facilities. Due to Du Pont's attempts at consolidating its hold on the industry, both the Army and the Navy – some of Du Pont's most important customers – had grown suspicious of Du Pont's anti-competitive activities. Francis argued the need to have research done by experts, so as to increase products as well as soothe relations with the government, and display Du Pont's "progressiveness" (ibid., 33). The Executive Committee approved, and the original mission of the General Experimental Laboratory – later renamed the Experimental Station – broadened again.

The boundaries drawn between both laboratories would soon face another challenge, in 1904, when a subcommittee was tasked with investigating research management. The main questions tackled were the assessment of the differences between a departmental laboratory, such as Eastern, and the General Experimental Laboratory, as well as their rivalry. Although the success of the Eastern Laboratory was generally acknowledged, the department heads for smokeless and black powder recognized no need to start their own departmental laboratories, instead supporting Francis' plan of a general, centralized laboratory. In turn, Barksdale argued that Eastern's success was in part due to its association with a plant and providing access to raw materials as well as training for plant managers. He further emphasized that the question should not be whether to do research or not, but rather how to utilize research expenditures efficiently (ibid., 28). In the end, this debate was not resolved by the questions regarding what kind of scientific work was being done at the laboratories, but rather the costs and appeals to business sense. It was decided essentially to embrace both decentralization and centralization, keeping the central and departmental laboratories (ibid., 29).

Francis I. du Pont, a Yale graduate in chemistry and inventor who had gathered practical experience as a plant superintendent, was hired as the Experimental Station's first director. Although, as Hounshell and Smith (1988, 35) point out, he "fitted the mold of lone inventor far better than that of credentialed chemist", turning out to be a rather poor research manager.

From WWI to "Fundamental Research": The Laboratories' Early Decades

The early years of both laboratories are characterized by growth and expansion, but also managerial struggles. When a financial panic hit in 1907, leading to layoffs at the Station, an Experimental Board was founded to oversee the Station's work, which had fallen into disarray due to Francis' leadership. Allocating research resources and instituting sounder management practices, the Station soon found itself under the leadership of Pierre S. du Pont, Barton, and the Experimental Board, as well as the head of the Development Department, Irénée du Pont. Fin Sparre, a Norwegian chemist and engineer with academic training in Germany, was promoted to chief chemist. He expanded the Station's staff, employing 35 chemists by 1911, many with "excellent academic credentials" (Hounshell & Smith 1988, 49). The Station's "long shot" (ibid., 47) kind of research proved hard to quantify, as evidenced in 1910, when the Executive Committee requested a report to find out whether the Station "paid its way". Irénée struggled to come up with any kind of supporting numbers, yet the Executive Committee exhibited a continued commitment to research.

Compared to the Experimental Station, the Eastern Laboratory's clearly defined mission and successful work on product and process improvements allowed it to generate impressive profits. Reese was eventually promoted to director of the new Chemical Division of the High Explosives Operating Department, leaving Arthur M. Comey in charge. Comey, having done graduate work in Heidelberg, Germany, had served as a professor at Harvard and Tufts as well as a consulting chemist. His splendid reputation helped attract academic chemists to the laboratory (ibid., 52). By 1911, the Eastern Laboratory's staff numbered about 80 employees, 27 of which were academically trained chemists, many boasting PhDs from German universities (ibid., 53).

In the following years, the organizational form of the laboratories was directly influenced by corporate strategy. This became evident when reorganization toward greater centralization was undertaken in 1911, leading to the creation of the Chemical Department and putting Reese in charge of both the Eastern Laboratory and the Experimental Station (ibid., 58). Once again, quantifying the value of research became a problem, with Reese arguing for a different kind of profit to be produced by the Station: a good reputation and relationship with the Army and the Navy. Yet Reese not only had to fend off inquiry from the Executive Committee but also attacks from the Department Heads, who, in the move to centralize R&D, felt Du Pont's research organization had become unresponsive to their practical problems (ibid., 56) and so lobbied for the creation of departmental laboratories. In the wake of structural changes following an antitrust suit in 1912, the Executive Committee was reconstituted in 1914. Eventually, in 1915, E.I. Du Pont de Nemours & Company was incorporated and it purchased all assets from the previous Du Pont Powder Company, becoming the sole parent company.

The outbreak of WWI in Europe led to great transformations that manifested in two areas: a rapid increase in demand for smokeless powder, as well as a dyestuffs shortage due to the British blockade of German-produced products. Du Pont reacted by rapidly expanding smokeless powder plant capacity, also by moving into the dyestuffs business – a decision justified not only by strategy but also political necessity (Chandler 1990a, 85f; Hounshell & Smith 1988, 77). Sparre headed the diversification efforts. Tensions between managers and researchers arose: The lesson learned by management from the move into dyestuffs was to buy firms and improve their products through their own R&D program since the venture into dyestuffs had a slow and straining start and the Chemical Department did not prove responsive enough. In contrast, researchers had wanted to "home-grow" the dyestuffs business, and convinced management to do so with initially poor results. In the end, this episode leads to managers being suspicious of researchers' opinion in the years to come (Hounshell & Smith 1988, 77, 97). During those years, Du Pont looked not only inward to its R&D organization for the tasks at hand but also outward, seeking to build relationships with universities to ensure a continuous stream of the best graduates and professors to act as consultants, as well as the latest scientific news (Steen 2014, 261; Rhees 1987, 71). From 1918 onward, Du Pont funded fellowship programs for graduate research at prestigious universities – among them Harvard and MIT.

The 1919–1921 postwar recession hit the company hard, causing losses especially for the dyestuffs department (Chandler 1990a, 104). By 1921, the Executive Committee had to react, responding to the rough situation by adopting a multidivisional, decentralized structure, with the individual departments of explosives, dyestuffs, etc. run as separate businesses (ibid., 106ff; Chandler 2005, 43ff). This ultimately leads to the Chemical Department's decentralization – its inflexibility during war times being one of the major causes, and the departmental managers desire for divisional laboratories another (Hounshell & Smith 1988, 107f) – which left the Chemical Department "dismembered" (ibid., 109). As the newly allotted budget was only a fraction of its prior size, the remaining group of scientists was tasked with general chemical research. Still, the divisional squabble for control over research illustrates just how ingrained the culture of R&D at Du Pont was at the beginning of the 1920s, with the divisional managers arguing: "... the improvement of our products through research ... are so vital to most of our industries as to be the deciding factor in determining success or failure" (Patterson, quoted in Hounshell & Smith 1988, 108). The reorganization set the stage for Du Pont to become a company based in a broad range of chemical industries. Yet the company did not conquer the market through R&D, but rather through the acquisition of competitors and patents – the departmental research laboratories being used to fit new technologies and products to emerging markets, which was led by Sparre and the Development Department (ibid., 119).

Why was the Chemical Department not abolished completely, but instead transformed into a central research organization? For one, while closer to the needs of the manufacturing divisions, the highly specialized research at the departmental laboratories did not allow for general or long-range projects. Second, the Executive Committee's decisions were swayed by a strong proponent of research: Charles M. A. Stine (Hounshell & Smith 1988, 125). Stine joined the company in 1907 as a chemist in the Eastern Laboratory (ibid., 51), later ascending to supervise the organic research work of both laboratories (ibid., 86), and ultimately becoming assistant director of research for the Chemical Department in 1919 where he actively spread his "enthusiasm for science" (ibid., 126). But it was only after Reese stepped down as research director in 1924 that Stine managed to convince the Executive Committee that a science-oriented central research organization was needed. After a failed attempt at doing joint research with General Motors (ibid., 126ff; cf. Chandler & Salsbury 1971, 433ff), Stine shifted strategy, exploiting the growing diversification of the company by building alliances with the departmental laboratories, since almost all of the Chemical Department's budget came from doing contract research for the industrial departments. Stine successfully argued that the Chemical Department needed to stay in touch with the departmental laboratories on the one hand, but on the other would need a large increase in money and manpower to come up with anything big (Hounshell & Smith 1988, 135f).

No discussion of the early decades of Du Pont's R&D efforts would be complete without mentioning "fundamental research". It all started with a memorandum by Stine sent to the Executive Committee in 1926, titled "Pure Science Work". In it, Stine advocated for undertaking "fundamental research work" (ibid., 223) without products or processes in mind, but rather the discovery of new scientific facts. These new facts would serve as the "raw materials" (ibid.) of applied research, a metaphor he reinforced by conjuring Herbert Hoover's image of the depleting reserve of scientific knowledge (Hoover 1926). He further referred to the successful examples of such projects in Germany and at GE to substantiate his claims (Hounshell & Smith 1988, 223; Sturchio 1984). Four reasons were listed to undertake this new venture: First, it would bring prestige not only in academia but also in the eyes of the general public by shaking off the company's "merchant of death" image (Cerveaux 2013, 267).[6] Second, purely scientific research would improve morale and aid in recruiting. Third, results could be used as bartering chips when trying to get information from other research organizations. Fourth, and last, new products or applications might come forth from the project (Hounshell & Smith 1988, 223; Sturchio 1984). In Stine's detailed proposal of 1927, after consulting with – among others – Whitney of GE, he dropped the notion of "pure science work" for the label "fundamental research", although it is not entirely clear as to why he switched.[7] Stine argued that, while a gamble, the new program could aid in discovering principles that would allow materials to be approached in a more scientific way. The Executive Committee passed the proposal, probably due in large part to Stine's "salesmanship" (Hounshell & Smith 1988, 226). Early in 1928, Stine hired Wallace H. Carothers, and around him a group of PhDs from prestigious universities such as MIT, Princeton, Harvard, and Johns Hopkins. The brilliant Carothers would lead the research team that produced the breakthroughs that Du Pont R&D is most famous for: neoprene and nylon (ibid., 229; Smith &

Hounshell 1985). But commercial success with nylon only came after Stine yielded his position as research director to Elmer Bolton, moving up to the Executive Committee himself. Bolton, a "pragmatic research director" (Hounshell & Smith 1988, 237) reorganized the Department after the Great Depression and pushed Carothers' group toward the development of nylon as a commercial fiber. It was these successes that ensured Du Pont R&D – and the label "fundamental research" – places in history.

1907: AT&T

The Bell Telephone Laboratories, incorporated in 1925 as AT&T's research organization, quickly became one of the most famous and most visible industrial R&D laboratories (Reich 1980, 506f; Lipartito 2009, 132). To understand where the eponymous Bell Labs came from, a similar approach to what has been discussed regarding GE and Du Pont will be taken: First, the organizational framework and its struggles in a rapidly changing telecommunications industry will be explored, before shifting perspective toward the detailed happenings within AT&T, then to the establishment of the company's first laboratory which ultimately became Bell Labs.

Bell Telephone and the Business of Telephony

The organizational history of Bell Labs begins with Alexander Graham Bell's work on harmonic telegraphs in the 1870s and the invention of the telephone, which patents were granted for in 1876 and 1877. After being turned down by Western Union (WU), the dominant communications company of the time, Bell and his partners organized the Bell Telephone Company in 1877, soon becoming a successful business by renting telephones, which elicited competition in the form of a WU subsidiary, the American Speaking Telephone Company (Lipartito 2009, 133). In a 1878 reorganization of the Bell Telephone company, Theodore N. Vail, the previous manager of the Railway Mail Service Commission, became general manager. Renowned financier Jay Gould, owner of the American Union Telegraph, planned to purchase Bell and combine his companies into a telephone-telegraph system, a direct threat to WU, which in turn decided to strengthen Bell to fend off Gould. WU sold all of their telephone-related patent rights to Bell, including rights to long-distance telephony, in 1879 (ibid., 134). Consolidation and another bout of reorganization followed, and the American Bell Telephone Company (ABT) was born in 1880.

The 1880s were characterized by moves to build a monopoly. In an effort to increase manufacturing capacities, the growing ABT company purchased the Western Electric Company of Chicago in 1882 in order to act as their manufacturing arm. Competition was fended off by continuous purchasing of patents, ensuring not only a sound technological basis, but also high barriers for entry into the market for any inventor who may have wanted to capitalize on an invention (Reich 1985, 137). To focus on long-distance telephones a subsidiary was founded: the American Telephone and Telegraph Company (AT&T)[8], which took over as the parent company of the Bell System in 1899 (ibid., 139). But even the steep gain of customers and major growth could not stop an imminent power struggle that started when a banking syndicate, led by New York financier J. P. Morgan, purchased a massive share of AT&T stock in 1902. Morgan used his leverage to bring back his confidant Vail, who had left the company in 1887 to pursue his own business ventures. The Panic of 1907 was the final straw that tipped the balance of executive control of AT&T into Morgan's favor, with Vail as the new president following the resignation of his predecessor and parts of the board of directors. The pivotal year of 1907 brought not only great transformations for the future of the telephone business, but also how AT&T would handle science and invention.

Before the managerial and structural reshuffling that brought about Bell Labs is discussed, it is imperative to understand where product and process improvements at AT&T and its predecessor companies originated from. As several authors note (Reich 1985, 142; Wasserman 1985, 35), AT&T's main area of revenue – telephones – were put on a shaky technological basis from the start, as well as the impending threat of the expiration of the underlying patents in 1893/94 (Reich 1977a, 20). Technical functions such as improvements and testing were first performed by Thomas Watson, minor experimentation being done by new hires. In 1880 William Jacques, one of America's first science PhDs, having received his honors in Johns Hopkins' physics/electrical engineering degree program, joined the company (Reich 1985, 143). He became head of AT&T's first formal laboratory, the Electrical and Patent Department in 1881. The department was under the direct oversight of Vail who was a strong proponent of its services. Yet in his stint at AT&T Jacques seems not to have influenced the company's research policy in any lasting manner (Hoddeson 1981, 520). In 1883 the Experimental Shop was founded as a "supplementary research organization" (Reich 1985, 143), and renamed the Mechanical Department in 1884. The department's director Hammond V. Hayes, boasting a PhD in physics from Harvard, was strongly supportive of scientific studies, especially since many contemporary technological problems were hard to solve without recourse to scientific basics, but when Vail quit in 1887 he was left without support on the managerial level (ibid., 144). Indeed, the Mechanical Department had to concentrate on daily problem-solving for other departments to such a degree, that Hayes soon scrapped any plans of doing independent research, in turn suggesting doing joint research with students at MIT and Harvard – a plea left unanswered by company president John Hudson.

Come 1893, the Mechanical Department was absorbed into the Engineers Department under the direction of chief engineer Joseph P. Davis (Reich 1985, 145). Hayes still managed to retain a certain degree of autonomy, with the Mechanical Department now focusing on the development of new apparatuses. At that time, 19 engineers were employed at the Department, few of whom had any advanced degrees, and most having started at the company as technical assistants and worked their way up (ibid., 147). The Department grew rapidly, as the staff increased to 148 men with advanced training, among them Frank Jewett, a physics PhD from the University of Chicago with teaching experience at MIT.

By 1905, the department was placed under Hayes' sole leadership as chief engineer (ibid., 148; Reich 1977a, 20). But interestingly enough, Hayes proved not to be very assertive of in-house research (Hoddeson 1981, 527). In his reports, he deemphasized the role of original invention and scientists, even in the face of successful research projects such as the loading coil, stating that the department would instead focus on incrementally perfecting existing technology, and that employing "men of unusual scientific attainment" would be an "expensive and probably unproductive undertaking" (Vail, quoted in in Reich 1985, 149). Hoddeson (1981, 527f) theorizes that Hayes simply responded to managerial attitudes that were skeptical of the values of science and the place of research, while Reich notes a general conflation of the roles of inventors, scientists, and engineers symptomatic of American culture at the turn of the century (Reich 1985, 149; see also Reich 1977a, 19). As a consequence of this research policy, the company had to rely on the incremental increase of purchased patents from outside inventors, a strategy that proved mostly successful due to AT&T's strong patent position (Wasserman 1985, 42). But before the transformations of 1907, the need to reconsider product innovation, and the coordination of design and manufacturing, became apparent (ibid., 38).

Establishing the Laboratory

The birth of AT&T's research organization occurred in the two pivotal years of 1907 and 1911, with 1907 bringing a reversal in corporate policy, and 1911 heralding the establishment of the company's first official research branch (Reich 1977b, 212). The 1907 Panic brought not only a reshuffling of AT&T's leadership but also severes cuts to staff and the demand to keep spending to a minimum. To reduce expenses, Vail consolidated the company's engineering in New York at Western Electric (Reich 1985, 151). After Hayes' retirement, John J. Carty ascended to the leadership of the newly consolidated research effort. Carty, having no scientific training, started at the company as a switchboard operator, receiving training-on-the-job during his time at Bell. In the meantime, Vail started to build, extend, and protect a "truly integrated national system" under his famous motto: "One Policy, One System, Universal Service" (ibid., 152f). With AT&T buying many of its smaller competitors in the years 1907–1913, Vail even anticipated federal involvement, cultivating a relaxed attitude toward regulation and attempting to build a benevolent relationship with the government for the Bell System (ibid., 154).

Since the Bell System was based entirely on wires, the newly emerging technology of "wireless" – radio – was starting to become a huge threat to AT&T's core business. In 1907, drawing on reports from Carty and chief patent counsel Thomas Lockwood (ibid., 155), Vail declined to buy radio patents. Yet awareness of the threat posed by radio grew, and both Vail and Carty decided to keep abreast of current developments. By 1909, radio had become a "very real" threat, especially to AT&T's stockholders who inquired about possible consequences (ibid., 157). Carty focused work on the development of an electronic repeater to boost signals and succeeded in convincing Vail that this repeater would be the solution for their troubles with long-distance telephony, as well as for controlling radio. In turn, Vail donned his broad support for a research program, approving laboratory investigations (ibid., 157f). Carty proved to be a splendid spokesman for scientific research, convinced that science would lead to "mastery of the forces of nature" (Carty, quoted in Hoddeson 1981, 529), in what can be seen as a distinct rhetorical shift from Hayes' directorship to Carty's (Hoddeson 1981, 530). He even went so far as to announce publicly that a transcontinental telephone line would be a reality by 1914, a bold claim that boosted the company's reputation.

For about a year, Carty and Jewett's research failed, which lead Jewett to studying possible avenues of inquiry and what kind of staff would be needed for such work, coming to the conclusion that scientists with advanced degrees were needed to investigate "areas of the physical sciences germane to communications technology" (Reich 1980, 512). Bringing this proposal to the board of directors, the creation of a small research laboratory within the Western Electric Engineering Department was approved and integrated into the branch organization: the Research Branch was born, and it was tasked with doing the "highest grade laboratory work" (Hoddeson 1977, 24). As both Vail and Harry Thayer, the president of Western Electric, were very technologically oriented and sympathetic to the endeavor, both boards of directors expressed strong support (Reich 1985, 159). The transformation of AT&T's approach to R&D was now complete: as a 1911 report stated, the cut-and-try-era was over, and "fundamental principles" were being investigated (ibid., 160).

Bell Labs: The "Idea Factory"

The new laboratory grew quickly in personnel and allotments: increasing from 20 members in 1912 to about 45 in 1914 and 106 in 1916, while the budget increased by a factor of three (Reich 1985, 176; Hoddeson 1981, 534). Even though cutbacks were made in other parts of the company due to an impending antitrust lawsuit, the Research Branch – and branches working closely with it, such as Development and Transmission – got increased support, having become even more attractive in the wake of the antitrust investigation. Research on the Audion triode lead to the successful inauguration of the transcontinental telephone line in early 1915, making Carty's bold promise come true. For this project, the Research Branch had not only worked on the theory and application necessary for boosting signals but also developed manufacturing methods.

In those years, Carty served as a connector between the laboratory and the board of directors, both lauding the laboratory's achievements and channeling management's expectations to the researchers (Reich 1985, 167). Having only limited scientific capabilities himself, he personified AT&T's research effort. Jewett oversaw the Research Branch and concerned himself mainly with recruiting and coordination, while the actual overseeing and directing of research was done by H. D. Arnold, a former student of the famous physicist Robert Milikan (ibid., 168). They all "took every opportunity to demonstrate that it [the Research Branch] contributed significantly to the Bell System's welfare" (ibid., 169), legitimizing growing expenses for research by referencing their highly conspicuous achievements. The laboratory's gospel was also used in a highly strategic way in shaping the company's public perception: For example, radio was depicted as a means to reach inaccessible places only, showcasing it as a technology that could not subvert the telephone, which kept subscribers and stockholders at bay (ibid., 175).

The war brought several organizational changes to the laboratory: Carty went to serve in the Army Signals Corps, and Jewett and Arnold were promoted to chief engineer and director of research, respectively. The Research Branch staff developed radio for the war effort and instructed military personnel on radio equipment use. When the war ended, Carty went on to become vice-president in charge of R&D at AT&T (ibid., 180), a position that allowed him to remain an important spokesman for research. By 1919, the growth of the Engineering Department called for another reorganization, which transformed the Research Branch into the "Department of Physical Research Engineering" (Reich 1985, 181). Its scope continued to widen after a decision to diversify research, while the accumulation of telecommunications-related patents served to protect core markets (ibid.). With Jewett at the helm, restrictions on interactions with outside scientists were relaxed. Prior to WWI, attending but not presenting at conferences and other meetings was the norm. Jewett cultivated a somewhat academic atmosphere at the laboratory, encouraging his staff to join scientific or technical organizations, yet keeping certain restrictions on publishing (ibid., 191ff; see also Lipartito 2009, 138ff). The interdisciplinary teams of physicists, mathematicians, engineers, and technicians helped the company's lawyers build claims in patent cases, by surveying scientific literature and working on projects geared toward patentable inventions (Reich 1985, 198f). Interestingly, the Research Branch's methods and ideas had spread outwards to the other branches of the Engineering Department soon after its foundation, more deeply ingraining a culture of science in the company (ibid., 201).

Soon, management decided to make research itself into a corporate entity, in part for the company's public image, but also to find a home for long-range research projects (Wasserman 1985, 40). The Bell Telephone Laboratories were incorporated in 1925 and actively promoted as "a kind of national research laboratory, where the fruits of pure research benefited science, technology, and American economic interests" (Reich 1985, 184). Jewett was chosen as Bell Labs' president and Arnold its director of research (cf. Hoddeson 1980, 432, figure 1), and 3500 people were under their employ in Bell Labs' first year (Reich 1980, 525). With a budget of about $12 Million, Bell Labs quickly became the largest and best-known corporate laboratory in the United States (Carlson 2007, 62). The laboratory kept branching out and diversifying its research (cf. Hoddeson 1980; Russo 1981), with its forays into solid-state physics culminating in receiving the Nobel Prize for physics for the development of the transistor in 1956 (Hoddeson 1977; Carlson 2007, 61ff). With such highly visible successes, Bell Labs cemented its place in the public consciousness as the "Idea Factory" (Gertner 2012; Ross 1931).

1912: Eastman Kodak

The last of the pioneers is Eastman Kodak. Compared to the previously discussed companies, there is considerably less scholarship on the topic (Dennis 1987, 486), as the Kodak Research Laboratory lacks the historical scrutiny that the other three went through regarding its genesis, staff, and practices. As will be seen, the story of Kodak's laboratory, and Kodak's organizational history itself, is less a struggle of coalitions of actors vying for resources and interpretative dominion, but rather a narrative of the vision and leadership of a strong company founder. This section follows a similar format: first, the business and technological environment and challenges of the company will be explored and the relevant actors introduced, before shifting to the birth of the laboratory itself and its first two decades leading up to 1930.

George Eastman, Kodak, and the History of Photography

Unlike GE and AT&T, whose early corporate history is characterized by a series of mergers and a various cast of actors, the history of Eastman Kodak[9] is inextricably tied to one man: its founder, George Eastman. Born in 1854, and having no scientific background, Eastman soon started to experiment with photography, filing a patent for a plate-coating machine in 1879 (Collins 1990, 38). The following year, Eastman started his own dry-coating business, and in 1881 formed a partnership with the businessman Henry A. Strong to form the Eastman Dry Plate Company (ibid., 42f). After the successful manufacture of dry-plate cameras, Eastman wanted to move into smaller "hand cameras" (ibid., 54), with the company soon debuting the first Kodak camera in 1888 (Coe 1976, 50ff). The novelty trademark[10] found its way into the company name in 1892 when the Eastman Kodak Company of New York was organized. During all those years, Eastman's initial partner, Strong, served as the company's president. Rochester, NY, home of Kodak Park, soon became known as the photography capital of America.

Although the initial founding and growth of Kodak as an organization was a rather straightforward affair, the business of photography proved turbulent, specifically due to two major innovations that reshaped the American photographic market in the late nineteenth century: the roll film system, and gelatin emulsions for developing film (Jenkins 1975, 176). The introduction of the roll film system created a mass market for photographic materials and apparatuses between 1889 and 1894, rapidly expanding the young industry until roughly 1909. Kodak soon found itself in a dominant position (Jenkins 1975, 178, table 8.1). To defend this powerful position, Kodak pursued a strategy of both horizontal and vertical integration, while a powerful patent position was built. Moves of horizontal integration can be seen in the acquisition of Boston Camera Manufacturing (acquired in 1895), American Camera Manufacturing (1898), and Blair Camera (1899).

Due to Kodak's commanding patent position in roll film cameras, as well as their technological troubles with any further development of the camera, many inventors chose to work on dry-plate cameras. Lots of small manufacturers began competing in a market that peaked around 1898, afterward characterized by intense price competition. In a move to broaden the company's base of products and improve focus on not only amateur photographers but also professional ones, Kodak moved into this market in 1900, purchasing many of the small manufacturers (ibid., 215). Instead of out-competing the many small competitors in technical innovation, Kodak chose to outperform them in marketing and supply (ibid., 192).[11] Interwoven with horizontal integration were moves to integrate vertically between 1890 and 1910: forward into distribution systems for the roll film camera, and backward into raw materials. Interestingly, Kodak chose to construct their own plants and manufacturing capacities instead of relying on acquisitions (ibid., 242ff; cf. Chandler 1977, 297).

George Eastman himself was a strong believer in "one man management", viewing the Board of Directors' value only as an "advisory instrument to a good manager" (Ackerman 1930, 98), and making many important decisions himself. The company was essentially still a partnership between Eastman and Strong, with the Board of Directors representing a "cross section of socially, culturally, and financially prominent Rochester families" (Jenkins 1975, 234) and the company structure remaining rather simple until the mid-1890s. Yet the following years can be seen as a gradual relinquishing of many of the functions Eastman initially performed himself, moving toward the construction of capacities for technological innovation within the company (ibid., 179).

One early pivotal figure at Kodak was Darragh de Lancey, a mechanical engineer and Kodak's first MIT graduate, who supervised the material production, proving to be an excellent organizer and innovator (ibid., 180). In 1920, Eastman credited de Lancey with having "switched Kodak Park from the empirical to the scientific path" (George Eastman Correspondence, GECP, January 29, 1920). Using his connections to MIT, de Lancey hired Frank W. Lovejoy in 1894, a recent graduate in chemical engineering, brought in to oversee the crucial film support department (Jenkins 1975, 182). In the following years, de Lancey and Lovejoy led Kodak's development efforts, in close cooperation with Eastman, who played the part of advisor and assistant. Indeed, Eastman displayed a surprising attitude and commitment to science and research, as evidenced both by his words and action. In describing his company, he depicted how "special chemical and mechanical departments with a staff of skilled hands are maintained for *experimental* purposes in order to keep in advance of all demands for improvements in every branch of photography" (Eastman, quoted in Ackerman 1930, 145, emphasis in original). At the same time, he reached out to professors at MIT, Johns Hopkins, Columbia, and Cornell University to recruit graduates willing to work on photographic chemistry, while having a PhD chemist do contractual research work for the company (Jenkins 1975, 182).

Lovejoy soon found a home at Kodak's Experimental and Testing Department, founded in 1890, among a group of college-trained chemists. The laboratory became the center for development work related to the company's backward integration, exceeding its analytical tasks by also improving on products and processes (ibid., 185ff, 301ff). With the company conquering new markets in the early twentieth century, a growing group of university-level chemists worked not only under the auspices of de Lancey and Lovejoy but also in other parts of the company. Many of the plants, shops, and other organizational units under

Kodak's roof did their own developmental work due to organizational constraints or completely different practical problems posed by the production of a wide range of different products, efforts that went largely uncoordinated with the Testing Department. In fact, Kodak's staff proved to be a very heterogeneous group, including graduate-level scientists and engineers, as well as technicians, designers, and craftsmen, often retained from acquired competitors in an effort to cultivate a group of highly skilled and creative technical and supervisory personnel (ibid., 235, cf. 180, 219).

The Birth of the Kodak Research Laboratories

Until 1912, two strategies for innovation had gradually formed at Kodak: Mechanical issues were solved by relying on the skills and creativity of the in-house staff, whereas chemical problems were tackled by purchasing patents generated outside the firm, with the Experimental and Testing Department then working on adapting the new product or process to Kodak's existing technological base (Jenkins 1975, 304). But toward the 1910s, a threat emerged that troubled George Eastman and Kodak's existing arrangements for innovation: color photography. In 1910, Eastman had already established a color laboratory at Kodak Park following a European trip where he witnessed the competition, but this venture ultimately failed. On his next trip to Europe in 1911–1912, Eastman was again made aware of the many German, French, and British inventors and chemists working on color photography within chemical firms.

At this point, the historical literature on the topic likes to relate the telling anecdote of George Eastman visiting Bayer, one of the leading companies of the German chemical industry at the time (ibid., 305f; Hounshell 1996, 25; James 1990, 50; Eastman Kodak Company 1989, 5). During the conversation, he was asked – some sources indicate by Carl Duisberg, a famous German chemist – how many people he employed in his research facility. Embarrassed, Eastman made up a number and later remarked to Joseph T. Clarke, his European correspondent, that "if Bayer can afford a research laboratory, we can" (James 1990, 50).[12] Eastman consulted with Clarke, certain that the success of any research venture was strongly dependent on the man leading it (Jenkins 1975, 107, 306; James 1990, 50). Clarke suggested C. E. Kenneth Mees. Mees, having graduated from the University of London in 1903 and received his PhD in 1906, worked at the English dry plate company Wratten & Wainwright as a partner and managing director while doing a considerable amount of consulting work on the side (James 1990, 36ff). When Eastman made his offer, Mees saw it as an opportunity to do proper laboratory research on problems of photography, a matter he valued highly, especially since he saw that little basic research had been done on the photographic process in industry at all (ibid., 49). Agreements were made pertaining to the laboratory's basic operating principles. No developments of major commercial significance in the first ten years were to be expected of the laboratory, and it would be operating independently from the other departments. Mees would report directly and exclusively to Eastman and Lovejoy, who would allow the researchers to publish their results unless they ran contrary to business interests (Jenkins 1975, 308). Taking two of his former colleagues from Wratten & Wainwright to Rochester, Mees went on to build his laboratory, modeling it after two templates. His first guide was William Rintoul, who established a famous laboratory at Nobel Explosives Ltd. in England, and his other major influence was Willis Whitney, whom Mees had encountered on a visit to GE in 1912 and later cited as inspiration on how to run a laboratory (Mees 1956, 28).

Mees' Laboratory over the Years

Initially, the primary goal of the research laboratory was to gain a scientific understanding of photographic processes, coupled with the conviction that a deeper, more thorough understanding would lead to new or improved materials and processes that could be commercially beneficial to the company (Jenkins 1975, 310). To tackle these tasks, the laboratory was organized in different departments, covering not only physics and organic and photographic chemistry but also motion pictures and color and practical photography (Kodak Historical Collection, Series 7, fol. 10). These departments brought different people with widely varying perspectives and skillsets together. The laboratory officially opened its gates in 1913 with the completion of the laboratory building, staffed by 17 people, four of whom held PhDs (Kodak Historical Collection, Series 6, fol. 15). The laboratory grew quickly, employing 40 scientists, technicians and service personnel by 1915, 88 by 1920 and 142 by 1928, with research expenditures increasing about tenfold between 1913 and 1928 (Jenkins 1975, 312; cf. also Kodak Historical Collection, Series 6, fols. 15, 16). Mees instituted a conference system to encourage discussion, and soon the laboratory published its own Abstract Bulletin so its researchers could keep updated on current developments (Jenkins 1975, 311).

According to Mees, whose research interests were reflected in the early laboratory work, most of the laboratory's capacities were used for the development of new products and processes, and about one-third were used to gain a fundamental understanding of the photographic process (ibid., 312). The laboratory's scientists provided advice to Eastman and the board in matters of patent purchases (James 1990, 75). Antitrust became a concern for Kodak, due to its numerous acquisitions of competitors in the early twentieth century. The Taft administration's 1912 antitrust suit was ultimately dropped, because of changeovers in the administration, as well as a softening of attitudes regarding big business after the giants had used their intense efforts made during the war for political leverage and public relations.

As illustrated by this episode, WWI played a pivotal role in Kodak's organizational structure, which was also felt strongly in the laboratory. Eastman was an enthusiastic supporter of the United States' entry into the war, and much of Kodak's production capabilities were geared toward aiding the government. The laboratory was fully supportive as well, especially after Mees made an emphatic plea regarding the advantages that research could bring (James 1990, 79ff). Soon, the Science and Research Division of the Signal Corps worked closely with the laboratory, which contributed to areas such as aerial photography, camouflage, or colloidal fuel. The most important contribution proved to be in fine chemicals, which had developed into a crisis due to the embargo on German synthetic chemicals. The production of such chemicals was institutionalized after the war in the Department of Synthetic Organic Chemicals (Jenkins 1975, 313).

Following the war, Mees kept adding new departments, which coincided with a general broadening of the research being done, with organic and physical chemistry especially beginning to play a bigger role (ibid., 314). Decision-making and leadership structures for the whole company began gradually to change as well, as Eastman slowly transferred competencies to his successor Lovejoy. Up until and including the war, the board of directors had still reflected company ownership, making Kodak a highly centralized corporate entity. When Eastman's old partner Strong died in 1919, steps were taken to reform the organizational structure toward decentralization, a strategy that was acted upon in some areas, yet cut short when a multidivisional structure similar to Du Pont was proposed. Still, the 1919 reorganization created a functional structure (cf. ibid., 329, figure 14.10), and the board of directors was soon comprised by a new generation of leadership personalities. The laboratory was a secure and reputable department within the company by this time. Mees served on the board of directors since 1923, and many of his former employees held important positions in the company due to Mees' habit of transferring laboratory staff into leadership positions (ibid., 314). Mees pushed for further extension of the laboratory, strongly advocating for it as soon as 1927 (James 1990, 136ff). When he made requests two years later, he was initially turned down, 1929 having been a bad business year for Kodak, and only succeeded after repeated queries. This episode illustrates that even while research had found its way to the core of the Kodak organization, and Mees was deeply entrenched and connected, the ups and downs of business still influenced the research organization.

By the late 1920s, the laboratory had arguably become the world's leading photographic research institution (Jenkins 1975, 318). Mees was highly respected inside and outside of the company as a spokesman for industrial research, making his research philosophy widely known in a series of speeches and articles. It could be argued that he even managed to become the second most famous research director – after Whitney – being helped by his penning of an influential guide on the organization and management of industrial research (Mees 1920).

General Themes and Concepts

At this point, several similarities and differences in the pioneers' experiences of laboratory foundation can be seen, illustrating how the early spread of research laboratories was much more than a simple process of adoption. The explanations for the creation of these four laboratories, as synthesized from historic literature, can be found in Table 3.1. In the following, some key points shall be highlighted in the existing explanations for laboratory emergence.

In the literature on the GE research laboratory, it is generally acknowledged that the laboratory was founded as a result of competitive threats caused by GE's eroding patent position (Birr 1957, 30; Hounshell 1996, 21; Kline & Lassman 2005, 604; Reich 1985). In a move to insulate the company from technological uncertainty brought about by the rapidly changing market, choosing the laboratory as a solution was painted as the managerial strategy of the times. The rationale involved directly referred to the need to solidify market control through winning the patent race, and no implications at all were made regarding how the laboratory could transform the need to acquire competitors, a strategy that was made precarious under the new legislation. Indeed, with the Incandescent Lamp Manufacturers Association allowing for tight grips on the market, both European patents and the threat of market upsets due to disruptive new technologies that made old lamps obsolete seem to have been major motivators in the creation of the GE laboratory.

Especially in later historical literature on the topic, a strong Chandlerian perspective can be identified, as Dennis (1987, 486) notes in his deconstruction of these new "organizational histories". While important in helping us understand how the market and corporations intertwined to influence structures, and how managerial strategies manifested and reshaped the organization, such a perspective must remain incomplete since it takes for granted the emergence of the idea of the research laboratory (ibid., 487), as well as the symbolic struggles of translation that made implementing the idea possible in the way that can be witnessed in the case of GE.

Similar to GE, competitive threats are generally acknowledged as the main reason for Du Pont's need for research (Hounshell 1996, 24). But instead of other manufacturers – Du Pont controlled a large share of the explosives market and the pace of technological change seems to have

Table 3.1 Overview Pioneering Laboratories

Company	*Year*	*Laboratory Founded By*	*Main Reasons*
GE	1900	Coalition of actors: Steinmetz (chief engineer) Davis (patent attorney) Rice (vice-president) Thomson (inventor-entrepreneur, company founder)	• Defense against competitive threats due to eroding patent position • → Buying patents more expensive than own development work (Davis) • → Need to understand + discover new principles underlying patents (Thomson)
Du Pont	1902	Eastern Laboratory: managers of Repauno (Du Pont subsidiary), amongst them Barksdale & Haskell	Eastern Laboratory: • Defense against competitive threats • Improvement of product yields + quality • Keeping in touch with industry developments
	1903	Experimental Station: Experimental Committee led by Moxham and other members of Executive Committee	Experimental Station: • Product improvements • Improvement of relationship with government • PR: creating a "progressive" company image
AT&T	1907	Reversal of Research Policy: Vail (president)	Reversal of Research Policy: • Panic of 1907 • Vail's goal of building national system • Threat of radio technology
	1911	Establishment of Research Branch: Carty (research director) Vail (president) Thayer (president of Western Electric)	Establishment of Research Branch: • Threat of radio technology • Failure of previous research
Eastman Kodak	1912	Eastman (company founder, president)	• Competitive threat of color photography • "Fashion" of having a laboratory

been different than in the electrical industry – the company faced the government as the main competitor, in the form of Army and Navy embarking on their own explosives research venture. Moreover, Du Pont had to tread lightly facing anti-big-business sentiments and the possibility of antitrust suits, a fate that could not be averted in 1912. Regarding the organizational shape of the laboratories, it has to be noted that much of the work on the early Du Pont company has been done by Chandler himself or from a Chandlerian perspective (Chandler 1977; Chandler 1990a; Chandler & Salsbury 1971), emphasizing corporate transformations of the era under the tenet that organizational structure follows strategy. Thus, in this view, the establishment of the laboratories and their later reorganizations followed simply from the du Pont cousins' provisions to centralize and integrate vertically. The efforts to diversify during and after WWI led to the need for departmental laboratories that served as expert consultants for a varied portfolio of chemical products.

At AT&T, competitive threats played an important role as well (Hounshell 1996, 24). Yet many authors cite either increasing technological complexity or technological needs as the main factor that transformed research at AT&T (Carlson 1997, 216; Dennis 1987, 486; Hoddeson 1980, 23; Laporte 1983, 96; Reich 1985, 177). The mounting challenges posed by increasing the telephone's range proved too difficult for craftsmen and technicians, and staff that could investigate the underlying – that is, scientific – basis of the technology were needed. Lipartito (2009, 134f) bases his explanation on a somewhat related line of reasoning: The mitigation of risk caused by technological changes as the prime reason for research that would help reduce uncertainty. But a detailed analysis paints a more nuanced picture, as Reich (1985, 178f) notes: The successful example of GE's laboratory surely influenced AT&T's leadership, while the replacement of top management in 1907 brought a group to the head of the company that viewed organized, scientific research very favorably, underlining the role leadership played in the implementation of organized research (Lipartito 2009, 134). It is evident that the external shock provided by the 1907 Panic was imperative in laying the groundwork for a culture of research.

Securing the patent position also holds true for Kodak (cf. Carlson 1997, 216), with the emerging technology of color photography about to shake up the market. At the same time, while hardly playing a part in Jenkins' detailed coverage, Hounshell and Sturchio both mention the looming specter of antitrust investigations brought about by the Sherman Antitrust Act that was increasingly enforced in the new century's first decade (Hounshell 1996, 25f; Sturchio 1985, 8f), causing a decisive change in company strategy, by having innovation be generated within the firm. A last factor, connected to the anecdote related above, is what one might call a fear of missing out to the competition, especially in Europe, and a growing "fashion" of laboratory research in industry (Dennis 1987, 486; Sturchio 1985, 7). Eastman was most certainly aware of European laboratories, as illustrated by his alleged discussion with Duisberg, and realized that a laboratory employing scientists had become necessary not only to keep up with the competition but also to keep up appearances. This is further spelled out by several features of the laboratory, such as the permissive attitude toward publication that directly translated into the visibility of the laboratory's efforts. It stands to reason that Eastman also intended the laboratory to garner prestige for his company (James 1990, 50), building a bridge toward a seemingly growing appreciation for science in American culture as a whole – a highly interesting facet of the industrial laboratory's history that shall be explored further in this thesis.

As can be seen, for all four companies, competitive threats played the most important part in the executives' decisions to build laboratories, with antitrust reduced to a minor role. But the detailed histories allow for a much more nuanced understanding than simply relegating laboratory emergence to clear-cut managerial strategies of the time as the organizational histories might suggest. In fact, at GE, it took several attempts and a coalition of different actors within the company to convince management of the need for organized research, a move apparently not self-evident to managers. These actors used their respective positions to influence the Board of Directors, utilizing different legitimizing strategies: Davis mentioning the costs of falling behind the competition, Steinmetz outlining the practical implementation of such a venture, and Thomson stressing the shortcomings between departments and the need to investigate underlying principles. As Carlson (1995) shows, the strategy and structure of TH and therefore GE were strongly shaped by a variety of interest groups bargaining with each other, the birth of the laboratory being no different (cf. Dennis 1987, 487). Both the necessity of the laboratory to prove its own worth, as well as its standing and relations within the company, can also be understood through this lens (Reich 1985, 107f; Carlson 1995, 89).

At Du Pont, the different places in the company's structure where the Eastern Laboratory and the Experimental Station originated underscore the need for a detailed view. While the Eastern Laboratory came from the needs and experiences of practical work on the Repauno shop floors, the Experimental Station arose from Moxham, a member of the Executive Committee. Threatened by the new laboratory, the Eastern Laboratory's proponents managed to dramatically trim the Station's purview, and only external factors – Du Pont's relationship to the Navy – bolstered Moxham's claims, returning competencies to the Station. Similar struggles are evident in the following years and decades through reorganization and diversification, illustrating how tasks, duties, scientific freedoms and even organizational setting were sometimes shaped by outside intervention, but mostly by individual interests and the rivalry between the two laboratories that was made especially obvious in the attempts to quantify the value of research. Furthermore, the venture into "fundamental research" happened because of Stine's entrepreneurial actions: "selling" science to Du Pont's leadership, and drawing upon a coalition of supporters inside the company (the industrial departments and advertisers) and outside (experience by Whitney and others, as well as utilizing metaphors for scientific progress that were supplied by Hoover).

At AT&T, the move toward scientific research and ultimately Bell Labs may be seen as a gradual evolution of science and engineering in the Bell System (Hoddeson 1981, 542). The laboratory did not grow from scratch (Kline & Lassman 2005, 605) but was yet another branch of the Engineering Department, with much of its initial staff, and especially its leadership, already present at the company. From this vantage point, the place that the laboratory took in the organization for the most part – from the way research was structured to the tolerance that individual researchers were granted – can be understood as an organizational trajectory. It required individual attitudes toward science and taking advantage of managerial moods, to establish the organizational unit (cf. Lipartito 2009, 135ff). In Carty, we also find an actor playing the part of a "salesman" who transformed the promises of science into rhetoric that could persuade directors and decision-makers, whereas company president Vail was ready to shift AT&T toward a new path.

Contrary to GE, Du Pont, and AT&T, where top management had to be persuaded to make such a commitment and was expecting returns that justified the expenses, the Kodak Research Laboratory had a strong supporter in the company's highest office: George Eastman. Due to the tradition of technical innovation at Kodak as illustrated by the work of engineers and chemists since the 1880s, Eastman appreciated the support that scientifically trained men could give his company, which was articulated plainly by him crediting de Lancey for switching Kodak to a scientific path (Jenkins 1975, 180), and by his many supportive statements in regard to the scientific approach and funding (George Eastman Correspondence, GEC letters from June 21, 1910; July 9, 1920; December 21, 1921). With such a strong endorsement, it was possible for Mees to negotiate the policy of having no expectations for ten years, as mentioned earlier, when setting the groundwork for the laboratory. Thus, Eastman structurally occupied the roles that Steinmetz or Carty played at GE and AT&T, respectively.

One final question remains: Why were *scientific* laboratories established, and why were *scientists* hired? As evidenced before, depending on outside inventors for new products was standard proceedings for corporations at that time, and GE even had their own inventor, Thomson, working in-house. This shift in strategy is generally explained through reference to an increased output in scientific manpower on the one hand (Carlson 1997, 214), but on the other a change in perception of scientists vis-à-vis inventors, that is, a cultural transformation (ibid., 214f; cf. also Carlson 2007, 56). Inventors were said to work based on genius, such as Steinmetz (Friedel 1992, 23), depending on the crucial "eureka moment", whereas scientists projected predictability and continuity, "a language that made sense to managers struggling to protect big firms in the face of uncertainty" (Carlson 1997, 215). This can also be seen in the vague boundaries drawn between scientists, engineers, inventors, and so on (Reich 1985, 149; see also Reich 1977a, 19). Clearly, there was more to those roles than simply the functions they performed, especially since those tasks were often conflated as well, with the differences between invention, engineering and science being muddled. Here we can see early glimpses of the need to account for a discursive dimension that shaped and supplied cultural images of the inventor, the scientist, and so on, that informed actors' decisions in those days and allowed actors such as Steinmetz or Carty to build a legitimate, supportive base that would convince management to act upon their plans for laboratories.

Preliminary Findings from the Research Pioneers

In the following, some conclusions shall be drawn from the existing research on the four pioneering laboratories, especially regarding the cultural value of science. To conclude, the question of what made the four companies pioneers – their efforts themselves, or subsequent scholarship on them – shall be discussed.

General Explanatory Notions

Explanations for the birth of the industrial research laboratory can generally be grouped into two broad categories, differentiated by the explanatory mechanism. On the one hand, older literature often hinges on technological change, whereas newer studies from the late 1970s onward see organizational changes as the main factor that brought about corporate R&D.

Technological change as an explanatory mechanism focuses on the objects at the heart of the corporate endeavor, the things that were produced and sold: light bulbs, explosives, chemicals, cameras, and so on. From this perspective, these objects grew increasingly complex with time, and subsequently, it became increasingly complicated to make improvements to any of them, a precondition for staying competitive in an ever-changing, highly competitive marketplace (Bartlett 1941; Hall 1954; Birr 1966, 1979; Taylor 1976, 274). The methods and knowledge – craft knowledge gathered by years of experience, oftentimes characterized by a hands-on, cut and try, tinkering approach – utilized by those who worked on those objects no longer sufficed to tackle the problems posed by technological advancement. That is to say that the craftsman, the inventor, or the machine shop worker could no longer cope, thus industry turned to scientists who promised solutions for the problems facing corporations seeking stability. Implicit in this view is a fundamental difference between the knowledge types that the untrained or self-taught inventor used and the academic pool of knowledge with highly codified methods that were drawn upon by the professional scientist. Subsequently, a clear division is made between the scientist and the inventor, with the scientist seen as the next step in the corporate quest for innovation, thereby making the inventor obsolete. Moreover, another implicit assumption within this perspective is a teleology of technology: that technology necessarily grows increasingly complex with time.

This explanatory mechanism does not explain why the organizational entity of the R&D laboratory was placed inside the firm instead of contracting for it, a practice previously followed as outlined above: many companies relied on a variety of independents for patents, testing, and quality control. Then why hire a group of scientists and put them inside the firm? Second, this perspective struggles to explain why scientists were chosen at all. It hinges on the fundamental assumptions that methods and knowledge differentiate scientists from their competition and that a general supply of scientists was available for industrial work. But why did corporate leaders turn to scientists? Were the efficacy of science and the promise of the scientific method apparent and diffused throughout executives' offices to sway their decision-making? Had science built up such a rapport with the turn of the century that any corporate leader would want to spend large sums on a laboratory? And why, then, did the early laboratories have to prove themselves in the eyes of the managers?

Next to the assumptions regarding the nature of scientific and non-scientific knowledge, as well as the nature of technology, a problematic science-technology relationship is posited: that science could solve all technological challenges faced in the fields of chemistry and electrical engineering, with the application of the scientific method immediately leading to better technology. From a sociological perspective, such assertions need to be read with care since ideas about the nature and relationship of science and technology, as well as their practitioners, were strongly informed by ideologies of science, its uses, and its place in the society of those times. Furthermore, the lines drawn between the practitioners of science and technology seem clear and selective today, yet they were hotly contested in the late nineteenth century, with the professional identities of the scientist or the industrial scientist only just emerging, in part through the advent of industrial research. As outlined earlier, the inventor did not simply disappear, unable to cope with complex technology, but rather continued to play an important but shrinking role in corporate innovation. Lastly, as Smith notes, mid-century accounts were strongly informed by the popular images of industrial research: "scientists inventing new technologies in a systematic and predictable fashion" (Smith 1990, 122). Once again, it is evident that not only ideologies of the time when R&D laboratories were established played an important part in their advent but also today's understanding of these processes is heavily dependent on contemporary ideologies of science and progress. Accordingly, the simple substitution of inventor for scientists clearly motivated by technological concerns was far more complex than assumed, and blind to any transformations in the environment of the corporations that established laboratories, especially legislative changes. Of course, this criticism is far from disclaiming any change in the nature of the technological challenges facing the pioneers, especially since newer scholarship scrutinizing what happened within the laboratories – focusing on the technologies themselves – mapped the changing demands and problems (cf. Carlson 1997, 216; Dennis 1987, 486; Hoddeson 1980, 23; Laporte 1983, 96; Reich 1985, 177), while others, focusing on the inventors, show a distinct change in patenting activity and educational background (Lamoreaux & Sokoloff 2009, 61ff). Yet explaining the advent of industrial R&D through the factor of technological complexity alone falls far short of accounting for the complex interweaving between science, business, and politics at the time.

The organization sits at the heart of the second explanatory mechanism, often invoked in more recent literature. From this perspective, the R&D laboratory as an organizational entity was born once it corresponded with the dominant organizational logic of the time. This logic came about with the birth of big business: the large, centralized business organization featuring integrated structures, guided by elaborate managerial hierarchies, and taking advantage of massive economies of scale and scope. In addition to the structural prerequisites, big business also made available the resources – in the form of capital – necessary for costly undertakings such as in-house research (Carlson 1991, 286ff). The dominant organizational logic also prescribed the strategy of integration: forward into finished products, and backward into raw materials, one of these being knowledge (Dennis 1987, 487; Carlson 1997, 214). In this way, new inventions could be developed all the way from their inception to the finished product, ready for the market and escorted through every stage in-house. The large corporation at the turn of the century moving suppliers and wholesalers into its bounds can be understood by following the organizational logic of pulling insecurity inside the firm in order to be able to better control it. The sale of the finished product was now under the direct control of the corporation, as well as the supply and quality of the raw materials. Similarly, technological uncertainty brought about by new inventions and destructive innovation could supposedly be controlled by in-house researchers. The oftentimes related topics of antitrust and other legislation, as well as the complicated networks of patent rights and patent sharing agreements, need to be understood from this vantage point, as attempts at controlling external risk, with these field-level factors finding a response in organizational structures – such as the R&D laboratory.

Many of the later accounts of the early years of the pioneer laboratories are informed by this perspective (Reich 1985; Wise 1985). These two authors especially, but also the whole of the literature focusing on this explanatory mechanism, are heavily influenced by Chandler's work on organizational development (Dennis 1987, 486). While it is certainly a very powerful framework for understanding how and why business organizations changed and converged to similar forms of organizing, there are also some problems with adhering too strictly to this perspective. As the only analytical framework utilized to understand big businesses of the time, the result is too often predetermined, reduced to just a logical conclusion: The industrial research laboratory simply becomes "another exemplar of the line-and-division managerial structure to which Chandler sought to attribute the success of firms such as Standard Oil, General Electric, and DuPont" (Mirowski & Sent 2008, 645). Too many questions go unanswered when over-emphasizing this framework, such as the employment of scientists working in groups rather than the tried-and-true method of solitary inventors as discussed above, how managers willingly made such a risky gamble in the quest for corporate security, as well as why they did so in organizational fields that saw the adoption of very similar strategies and structures for large organizations. The similarities stopped here, and seemingly every corporation had to "re-invent" the corporate setting for organized science, instead of relying on proven templates.[13] Additionally, as the histories related above indicate, individual or rather entrepreneurial agency most certainly played a large role in how the pioneer laboratories came to be – a key factor that hardly plays any part in Chandler's framework.[14]

As Mirowski and Sent note, a more nuanced view is needed, "supplemented by legal and political considerations, which Chandler largely shunned" (Mirowski & Sent 2008, 645). One possible extension can be found in Fligstein's work on the transformation of corporate control in that era (Fligstein 1990, 2008). Fligstein stressed that not only environmental or field-level changes lead to transformed organizational structures but also added an ideological level with the so-called "conceptions of managerial control" as collective managerial styles emerging in organizational fields, which were highly dependent on and in interplay with the political context of business. After the first wave of antitrust legislation put limits on controlling competition via patents and pricing, the next conception of control emerged, where firms relied on size and economies of scale and scope to threaten potential competitors and hinder new market entries. Stabilizing the production process by controlling inputs and outputs became a major priority, through horizontal, and later vertical, integration. Growing corporations, due to their size, could then, ultimately, control the market and the competition through oligopolistic pricing strategies. The emergence of the industrial research laboratory coincided with this conception of control, as its rise resonated with attempts to stabilize production with better products and to supervise the creation and development of new products until they were market-ready. Fligstein's model follows the lines of the organizational logic outlined previously, but includes the complex interplay of business and politics into the way that managerial decisions are made.

All explanations put forward so far add valuable perspectives to explain the emergence of the industrial research laboratory in American manufacturing. While it is certainly a "combination of these factors – economic, institutional, technical and intellectual" (Sturchio 1981, 85f) that influenced the way the laboratory was conceived and the eventual shape it took in the large corporations, explanatory models focusing too strongly on only one facet of the whole picture fall short of accounting properly for the phenomenon, and often also lack proper analytical frameworks that go beyond that proposed by Chandler (cf. Mowery 1981, 42ff). As Mowery notes, a distinctly quantitative angle is lacking from all these contributions – granted, detailed case studies of the pioneers may not be expected to put forward any type of quantitative analysis, yet broader overviews do lack such a perspective, failing to add any layer of not only the spread of industrial research but also the development of large corporations and their respective industries. Thankfully, necessary groundwork is done by Mowery and others (Mowery 1983a, 1984; Mowery & Rosenberg 1989; Mowery & Teece 1996) to expand on what happened after the pioneers' laboratories were established.

Of course, one might ask why another book answering this question is needed. In regard to this study's conception as one of the diffusion of innovations, the detailed look at the pioneers reveals that the diffusion of the industrial laboratory cannot simply be treated as the spread of the innovation from one corporation to the next, who were compelled to establish their own organizational home for scientists due to isomorphic pressures. The situation can better be described as one of similar yet not identical conditions in the science-based industries (Smith 1990, 124) leading to the establishment and spread of the industrial research laboratory. Yet the laboratory itself took many different forms, heavily depending upon the actors who pushed for its realization, as well as the specific corporation and industry (cf. also Galambos 1979, 275). Thus, when analyzing the industrial laboratory, we must conceive of it as an instance of translation shaped by struggles for legitimation, and not as a fixed construct that diffused unchanged throughout a set of industries.

The Role of Agency

All the factors related above offer important insights into the changing environments of business, technology, and politics at the time. Needless to say, different explanatory levels feature varying degrees of fuzziness and necessarily fade out some parts of what happened. When drawing upon the very thorough accounts of the R&D pioneers, the importance of actors, and with them, agency, becomes instantly apparent. As the discussion of general themes and concepts regarding the pioneers has shown, these corporations can hardly be described as rational, decision-making machines when broken down to such a level. At each corporation, group of actors struggled for the dominance of their goals, each drawing upon different resources and building coalitions to bolster their agendas (Galambos 1984, 489f; Dennis 1987, 488; Carlson 1995), as evidenced at GE in the struggle to establish a laboratory in the first place, at Du Pont in the competition between the laboratories, and at AT&T in the new executive group's vying for power, just to name a few key instances. In a way, these actors functioned as translators – translating their goals and claims into a perspective that was both relevant and understandable for the companies' decision-makers (cf. Galambos 1984, 490), and emphasizing the complexity of the networks of power and decision-making of such organizations.

Glimpses from other corporations support this conclusion. The history of the Dow Chemical Company sounds reminiscent of Kodak since Herbert H. Dow was a strong company leader who valued the nature of scientific inquiry (Haynes 1939, 259ff). While there is no grand history written of the Dow laboratories, making it difficult to pinpoint when exactly the organizational unit existed, what can be found in ancillary works supports the hypothesis that the company was engaged in research from a very early point that was strongly supported by the company leadership (Whitehead 1968; Karpiuk 1984; Dalton 1995; Levenstein 1998). The story of Corning Glass further emphasizes the need to keep agency in view: the Houghton brothers, founders of Corning, experimented with various approaches to innovation, hiring their first professional scientist in 1904 (Carlson & Sammis 2009, 48), and in 1908 turned to their consulting scientist at Yale when the question arose of establishing a Chemical Department. The story of Corning illustrates once again the gradual evolution of corporate R&D, as well as the importance of the various approaches to innovation experimented with by the company's founders to bring about a research laboratory (Carlson & Sammis 2009). The experience of the Aluminum Company of America (Alcoa) underscores how coalitions of actors pushing for change may face insurmountable roadblocks. While boasting three laboratories in the first decade of the twentieth century that were tasked with analysis and testing, company founder Hall remained in a commanding position of their technical direction (Graham & Pruitt 1990, 67). Hall's negative experiences with in-house technological development and doubts about its profitability led to him standing in opposition to establishing a proper research laboratory, which was presented to him as fulfilling the functions of technical authority for various company branches, as well as delving into research on the company's core technologies (ibid., 73). Only in 1919, after Hall's death, and with competitive threats intensifying after their initial patents expired, was R&D put on a stable basis within the Alcoa organization (ibid., 101ff).

This key insight further underscores the role of translation, emphasizing the way actors can bring about change in organizational environments. Furthermore, the actors' various translations reframe their goals in a way that fits the companies' structures and situations, while ascribing to what Galambos (1984, 490) calls sources of authority that are drawn upon to bolster claims and elicit support. These sources of authority explain how the actor coalitions in these corporations substantiated their claims and pushed through their preferred agendas while weakening those of competitors fighting for a different course of action. Also, it is these sources of authority – as institutionalized resources for legitimation – that are of main interests to the questions posed in this book. The case histories related above strongly support drawing the conclusion that one source of authority became more and more important with the outgoing nineteenth century: science.

The Move toward Science

My argument is that the "scientification" of American society can help understand how actors within the pioneers pushed for their goal of organized, science-based laboratory research within the confines of the corporation, and how their arguments gained momentum and vigor in the eyes of the executives. This move toward science can analytically be disassembled into two components.

First, on the level of an organizational field, it means the growth of science in both manpower and organizations. Whereas in the mid-nineteenth century those who wanted to do research in science went to Germany to pursue their scientific interests, by 1900 the expansion of university capacities fueled by federal monies and philanthropy had led to an increased supply of scientists boasting advanced degrees, often at the PhD level, in the natural sciences (Kohler 1990). The demand for university professors and other academic positions did not keep up, effectuating an oversupply of scientific manpower – many of those ready to take up positions in industry (Carlson 1997, 214). But organizational growth also means the differentiation into a variety of sub-disciplines, the establishment of specialty journals and regular conferences as a means of scientific communication, the expansion of science publishing, as well as the foundation of organizations such as the ACS and the American Physical Society that not only focused their disciplinary activities but also allowed subgroups of scientists to act in a unified manner, giving their positions and policies a voice for the first time (Bates 1945; Shils 1979; Owens 1997; see also Reingold 1964; Kevles 1995). Yet only speaking of the growth of science in measurable terms could simply be added as a further layer to the organizational explanation above without accounting for the cultural, more indiscernible changes that took place. In fact, this would risk oversimplifying the phenomenon as a simple relation between supply and demand: industry had a demand, while science supplied the manpower, and the scientist in industry suddenly becomes a simple corollary. Furthermore, any explanatory model that posits the spheres of industry and science as completely separate societal subsystems, and sees the industrial laboratory as the result of a convergence of both spheres (cf. Sturchio 1981, 84ff; Taylor 1976, 274), risks conflating process with the result as well as locating agency at the wrong levels.

Which is why the second component needs to be added: the cultural attributes of the organizational field as expressed in discourse. This component is concerned with ideas about science, technology, innovation, and the source of expertise in the American society – called discourse in my theoretical framework. It is a consequence of the growth of scientific manpower and institutions, but also feeds back into them. Hints of this cultural shift can be seen in the gradual evolution, not revolution (Carlson & Sammis 2009, 38), from inventor to scientist as purported in the literature, reportedly fueling the slow reduction in significance of independent invention. According to Carlson, inventors were generally seen to possess unique personal knowledge and skills, oftentimes called "genius", with their basis for expertise being personal and their successes relying on unpredictable "Heureka"-moments (Carlson 1997, 214). In contrast, scientists were portrayed as working in a clear, methodical manner, making nature tamable, being able to predict the behavior of natural processes and as such reducing risk and unpredictability – those facets that business managers of the time attempted to minimize in their restructuring of big business (cf. ibid.; see also Hughes 1989, 138f). Enhancing the organizational perspective with a discursive one makes it possible to understand how the cultural shift resulted in the replacement of individual genius as the source of technological progress in the dominant discursive construction – genius had been replaced by expert knowledge (Friedel 1992, 19ff; Lucier 2009), with the myths of the sole inventor without regard for science and the research scientist attacking problems in systematic ways now battling for symbolic hegemony (Reich 1985, 145f; Reich 1987, 344; Carlson 1997, 208). The organizational makeup of the field influenced the discursive battles, since, as Hintz notes, scientists had strong "pro-research associations like the American Chemical Society" (Hintz 2011, 737) and later governmental agencies, such as the NRC (post-WWI), on their side, which led to their side being able to afford "far more ink than the independents, and the public was left with only one side of the story" (ibid., 738). Independent inventors, lacking any umbrella organizations, thus lost the symbolic battle. This discursive evolution was further strengthened with science boosterism during WWI, which brought the "gospel of industrial research" (Rhees 1987, 15ff) to the fore.

Of course, framing the emergence of organized R&D as concurrent with a cultural shift is not entirely new and has been hinted at by a variety of authors, yet it has not gotten enough attention. While Carlson speaks of the "cultural efficacy of science" (Carlson 2007, 69; see also Carlson 1991, 347; Carlson 1997, 214), Shils notes that "larger movements of society where favourable" to the goals of scientists (Shils 1979, 35, also 30), Birr speaks of a "felt need for science" (Birr 1979, 179), and Reich of a "progressive spirit" (Reich 1977a, 16), whereas Sturchio refers to a less abstract "rhetoric of utility" that influenced science and industry at the time (Sturchio 1981, 84). But these brief glimpses illustrate how societal discourses on innovation and expertise saw fundamental changes from the Gilded Age to the postwar era, and I argue that precisely the discourse of the organizational field of corporate R&D served as a source of authority for actors within the pioneering corporations, legitimizing their ambitious plans for the firms' futures, while also needing to be translated to fit corporate objectives and the managerial worldview.

It is precisely this cultural, discursive dimension that has not been decidedly included in any analytical framework attempting to draw theoretical conclusions from the advent of industrial research. This shall be done in the course of this book – framing the spread of the industrial research laboratory as a case of diffusion in an organizational field. The historical literature offers hints at where to look when mapping the field's discourse and especially the crucial shift in discourse that supposedly happened around this time. The goal of this book is not to access and reconstruct the exact wordings and rhetorical strategies of argumentation that actors such as Steinmetz or Carty drew upon, since this work was thoroughly done by historians such as Reich, Wise, and Hounshell and Smith, but rather to look at the other side of the equation: the constructions they drew upon as informed by the discourse on science, the scientist, the professional, the laboratory, and so on. These fickle terms being constructed in discourses that are far more complex than plain references to notions such as utility and the taming of nature can broach, especially considering the symbolic battles fought for science and its applications within the scientific community at the time. What did practicing science at the time mean, what promises and claims did the scientific community make regarding the abilities of the scientific method, and what roles were discursively constructed for (industrial) scientists? Systematically charting the field's discourses, as textual expression of the institutions that were regulating organizational reality, will serve to complement the historical perspectives on the origins of organized research in industry.

The Status of the Pioneers

One question remains that, while not central to this analysis, will hopefully be further elucidated in the process of the discourse analysis. The designation of the laboratories of GE, Du Pont, AT&T, and Kodak as "pioneers" needs to be questioned with regard to whether this is due to actual pioneering efforts, or organizational and historical story-telling that raised them to the status of pioneers through an abundance of literature. The detailed descriptions of those four laboratories show that most certainly they were among the first to establish industrial research laboratories, and did so with hardly any or no templates to draw on, sounding out on their own the most productive and efficient ways of having scientists work in an industrial setting, often through trial-and-error. But within the literature, there is often mention of other corporations that early on used scientific methods to their gain. Standard Oil is sometimes reported as having organized research from 1890 onward (NRC 1940a, 259; Hounshell 1996 indicates 1906), Dow Chemical reportedly did laboratory research starting in 1897, 1900, or 1901 (NRC 1940a, 83; Hounshell 1996, 21; Sturchio 1981, 88), General Chemical in 1899 (Sturchio 1981, 88), Monsanto starting in 1904 (NRC 1940a, 191), and Goodyear in 1909 (Hounshell 1996, 21; Sturchio 1981, 88). If reported correctly, all these companies started their research ventures at the same time, if not earlier, than the pioneers. This raises the question of why their laboratories did not receive the same panegyric if any at all, possibly transforming our understanding of how the industrial laboratory came into being and how it spread.

The reasons for this can only be speculated. There is little systematic historiography on some of the companies mentioned, and while Standard Oil received a lot of attention due to its complex legal history (Larson 1969; Gibb & Knowlton 1976; Hidy & Hidy 1995), others received no historical scrutiny at all apart from self-published accounts of their own histories (General Chemical Company 1919; American Cyanamid Company 1937; Dow Chemical Company 1947, 1997; Union Carbide Corporation 1976).[15] One possible answer may concern the availability of company archives and the continuous existence of the company as a distinct unit: Some closed their gates or merged into other companies, while the four pioneers kept existing as – more or less – distinct organizations with a clear organizational core. Continued existence plays a big role, of course, in keeping archival material and making it available for researchers. Furthermore, each of the pioneers had certain characteristics that may have made them more interesting for historical scrutiny. Simplifying intensely, GE essentially grew out of Edison's workshop, Edison being one of the most venerated figures in American history. Researchers at Du Pont invented nylon and neoprene, outcomes of their ambitious program for fundamental research, and both products became ubiquitous in American society. AT&T connected the East and West Coast for the first time, while also doing groundbreaking and nobel-laureated research work on the transistor in its later years. Kodak, at last, boasted a storied founder in George Eastman, as well as a strong grip on the American unconscious with its marketing campaigns and pervasive cameras making any moment a Kodak moment. Of course, these suggestions are to be seen as highly speculative, tracing possible answers of how the pioneers got their status. To conclude, in the course of this analysis it needs to be remembered that the experiences of the pioneers were not the only experiences that we know of. Their histories are the most scrutinized and glimpses from other firms seem to confirm the broad environmental factors at play, as well as the need to better understand the surrounding discourse of science and its uses as sources of authority for those actors who pushed for change within the companies.

Notes

1 As Homburg outlined, the creation of the German industrial research laboratory was driven by two major factors: the *Reichspatentgesetz* (Patent law of the German Reich) of 1877, the first patent law of Germany that enacted patent protection for the first time, as well as creating a market for patents, and the growing number of academic chemists as along with interfirm competition that led to an increase in discoveries and inventions, thereby creating a need for in-house laboratories that could evaluate patents in order to remain competitive. As Homburg stresses, in addition to these main factors, "*local* factors" (Homburg 1992, 110, emphasis in original) differing between companies led to the different shapes and characteristics these laboratories took (Homburg 1992; see also Meyer-Thurow 1982; Hounshell 1996, 19ff; Murmann & Homburg 2001).

2 The 1921 survey is a "revised and enlarged" version of the 1920 one (NRC 1921, 1). All publications that utilize the NRC surveys use the 1921 survey.

3 Plenty of authors speak mainly of the big four when discussing the phenomenon of industrial research, these laboratories having become somewhat synonymous with early research efforts in industry (cf. Carlson 2007; Jenkins 1975, 305; Hounshell 1996, 21ff; Sturchio 1985; Birr 1979, 199; Smith 1990; Weart 1976).

4 In some accounts of the genesis of industrial R&D, Menlo Park is seen as the first R&D laboratory (cf. Rae 1979, 259; Israel 2002). Other authors rightly point to the differences between Menlo Park and later laboratories: Menlo Park was not integrated into the WU corporate structure and bears more similarity to R&D consultancies such as AD Little (cf. Carlson 1997, 207f; Friedel 1992, 19; Reich 1985, 43; Reich 1987).

5 As Dennis (1987, 488) notes, there is no textual or archival material available that confirms whether Steinmetz was aware of German corporate research laboratories, and he did not mention any German examples in his proposal to the Board of Directors.

6 When Du Pont failed to shake off the derogatory label, its advertising group instituted a radical change in the 1930s, birthing the now-famous "Better Things for Better Living … through Chemistry" advertising campaign. Stine's research

program helped shift the public image toward a science-based corporation, organizing exhibits such as "Wonder World of Chemistry" for the 1936 Texas Centennial Exposition (Cerveaux 2013, 267f; Sturchio 1981, 42; see also Rhees 1993b).

7 Hounshell & Smith (1988, 224) suspect that managers were more familiar with the term "fundamental research" than "pure science". The term "fundamental" certainly was part of the discourse on industrial research (Nutting 1917, 250). Both terms went on to be used synonymously in industry in the following decades (Cerveaux 2013, 269).

8 From here on, AT&T shall be used synonymously with any previous incarnation of the company unless noted otherwise.

9 From here on, unless necessary to denote a specific subsidiary or previous incarnation of the company, it shall simply be referred to as Kodak.

10 The trademark "Kodak" was coined by Eastman in 1888, chosen due to its brevity, easy pronunciation, and lack of association with any other companies or products in the photography business (Collins 1990, 55).

11 Indeed, Kodak's marketing efforts would go on to become both successful and famous, coining well-known slogans such as "You press the button, we do the rest", "Take a Kodak with You" (Coe 1976, 125ff; Collins 1990, 45ff, 157), and terms such as the Kodak Girl (Collins 1990, 156f) or the Kodak Moment (Munir & Phillips 2005).

12 The historical validity of this story is not clear since there is no definite source. Jenkins (1975, 306n15) makes a rather convincing case that it likely happened: Eastman was touring Europe during that time, and interviews with Mees corroborate that Eastman's decision for the laboratory was influenced by a visit to the Bayer chemical company.

13 In a perspective that relied so strongly on environmental factors bringing about a dominant organizational logic, which in turn influenced organizational structures, it is surprising how little of a role successful templates of the organization of research (the famous GE laboratory and the German example of successful research in the chemical industry) played in the early years of industrial research laboratories.

14 Of course, this lament is informed by works (e.g. Reich's and Wise's) using the Chandlerian framework and showing in detailed historical analyses the part that individuals played. My concern is not to disregard or object to their contributions, but rather to use their detailed insights and include them on a higher, analytical level for diffusion and organization theory on the one hand, as well as the historic case of the industrial research laboratory on the other.

15 There exist studies for several other companies focusing on research laboratories as central unit of analysis (Corning: Carlson & Sammis 2009; Alcoa: Graham & Pruitt 1990; Merck: Gortler 2000; Galambos & Sewell 1995; RCA: Graham 1986). They do help illustrate the lack of clear templates for laboratories, and how strongly driven by individual agency the establishment of laboratories was, but since this chapter focused on what the historical literature considers the "pioneers", they were not discussed in detail.

4 The Organizational Field: US Chemistry, 1870–1930

The organizational field of chemistry from 1870 to 1930 was chosen as research site to analyze the effects of discourse on the diffusion of innovations in organizational fields. The field of chemistry itself is defined here as those actors, organizations, and discourses in the environment of chemical firms establishing R&D efforts. That is, the organizational field comprises not only other chemical firms but also organizations from other industries, academia, and on the federal level. But before the field's layers can be unpacked, two questions need answering: Why chemistry, and why the years 1870 to 1930? In regard to the first question, as the detailed history of the pioneers showed, the two main industrial fields in which the early industrial research laboratories were established were related to and drawing upon either chemistry or electrical engineering: DuPont and Kodak can be grouped closer to chemistry, whereas GE and AT&T fall closer to the electrical engineering side. Of course, it can be argued that the latter firms' research involved chemistry as well, such as Langmuir's work at GE (cf. Taylor 1976, 274). Both fields, chemistry and electrical engineering, certainly possess their own identities and trajectories – electrical engineering in relation not only to engineering but also science, feeding from a multitude of influences such as physics and chemistry. In contrast, chemistry can be regarded as a scientific field that came into its own in the nineteenth century. While analyzing the discourse in relation to both fields would surely be fertile, charting the development of only one of those fields proved to be an extensive undertaking, since understanding an organizational field requires one to go farther than simple quantitative listings of actors, and more toward decoding of cultural changes and their relation to larger societal patterns and transformations. Thus, chemistry was chosen, and not least because the history of the discipline proves interesting, as the nineteenth century, sometimes called the "century of chemistry" (Nielsen & Štrbáňová 2008, 328), saw chemistry set out on a path of professionalization, differentiation, and popularization, culminating in the pivotal role it played during WWI.

As for the second question, the timeframe of 1870–1930, roughly thirty years prior to and after the birth of the first laboratories, was chosen since such a periodization allowed for assessing discourse before and during the establishment of the early laboratories, as well as changes within discourse in the following years. 1870 itself serves as a reasonable starting point since the transformation of chemistry into a national scientific discipline started roughly at that time, and was concurrent with important changes in the landscape of higher education such as the foundation of new universities. 1930 was chosen as terminus since it can reasonably be assumed that by then, industrial research laboratories had become firmly established within organizational discourse, with R&D laboratories having become "en vogue". WWI is often seen as a "watershed" moment in regard to attitudes toward science. By carrying on into the late 1920s, it is possible to see how discourse changed once normalcy had returned after WWI, and the United States' system of science funding, heavily modified during the war, was in place. Terminating in 1930 the great economic and political changes brought about by the Great Depression – causing layoffs in even the most successful R&D laboratories (Wise 1985, 283) – will not form part of the sample. Eventually, the goal of this chapter is to establish a timeline of important actors, events, texts, and topics, and to paint a multilayered picture of the development of the organizational field of chemistry in which the first R&D laboratories were founded.

Chemistry as Academic Science: Scientific Societies, Journals, and Chemical Education

Scientific Societies and Related Organizations

Before diving into the history of scientific societies that pertained to chemistry, let us take a brief look at the landscape of the most important general scientific societies in the United States (cf. Dupree 1976). The American Academy of Arts and Sciences was founded in 1780, and more than half a century later the American Association for the Advancement of Science (AAAS) was born in Philadelphia in 1848, with its stated goals being to promote scientific communication and foster collaboration. Membership in the AAAS was open to "all friends of science, whatever their actual attainments in research" (Reingold 1964, 200), with a fellowship system established to decorate the eminent scientists of the time. After postponing their meetings during the Civil War, the AAAS resumed activity in 1866 and grew considerably. The National Academy of Sciences (NAS) was founded in 1863 (Cochrane 1978; Dupree 1979). In contrast to the AAAS, the founders of the NAS aimed at creating a highly selective group of "leading scientific savants" (Reingold 1964, 200) that would come together to establish standards for the professional scientist, as well as seeking funding from the federal government. Together, the AAAS and the NAS became the two most important general scientific

Table 4.1 Foundations of Chemistry-Related Scientific Societies and Other Professional Associations, 1870–1930

Organization	*Year established*
AAAS Subsection C (Chemistry)	1874
American Chemical Society	1876
Association of Official Agricultural Chemists	1884
Chemical Society of Washington	1884
Chemists' Club	1898
American Society for Testing Materials	1898
American Ceramic Society	1899
New England Association of Chemistry Teachers	1899
New York Section, Verein Deutscher Chemiker	1900
American Electrochemical Society	1902
American Society of Biological Chemists	1906
American Institute of Chemical Engineers	1908
International Union of Pure and Applied Chemistry	1919
American Institute of Chemists	1923

Sources: Bolton (1902), Bates (1945), Beardsley (1964), Bogert (1908), Lovelace and Thomas (1919, 1920), Reese (1976a), Thackray et al. (1985), own research.

societies after the Civil War, shaping discourse and science policy in their wake.

Turning to chemistry, Table 4.1 readily illustrates that there were and are relatively few specialized societies since the increasing specialization of chemistry is covered by the elaborate divisional structure of the American Chemical Society (Thackray et al. 1985, 176). The "prehistory" of chemical societies starts in the late eighteenth century, with the birth of the Chemical Society of Philadelphia (1792), a local group of chemists that disappeared around 1809 with the death of its founder (Bogert 1908, 166f). Similar fates awaited the Columbian Chemical Society of Philadelphia (1811) and the Delaware Chemical and Geological Society (1821) (Bogert 1908, 167ff). The Committee of Chemists at the Convention of the Friends of Domestic Industry (1831) proved short-lived but important since it was the first time chemists as a group influenced policy, securing congressional legislation for preserving unprotected chemical industries against foreign interests (Browne & Weeks 1952, 8). As these cursory fragments show, while attempts were made at organizing within the chemical profession of America, they were still highly localized, as well as strongly dependent on their founders, whose deaths were often the reason for societies vanishing.

Within our timeframe of 1870–1930, the first proper organization of chemists can be found with the creation of a purely chemical branch of the AAAS, Subsection C: "Chemistry", a request granted in 1874. Yet both AAAS and NAS were often seen as lacking in key aspects, making them a far cry from a proper home for scientists, with the AAAS doing little between their meetings, and the NAS being described as "even more ineffectual" (Kevles 1979, 140). Thus, the ACS was founded through the recognized need for an independent society that would "unite in one active, aggressive organization the chemists of the country" (Bogert 1908, 170), as well as provide systems for indexing literature and setting standards for analytical methods (Beardsley 1964, 23). At the "Centennial of Chemistry" in 1874, the first steps toward establishing the organization were taken, and the ACS was officially founded in 1876 (Reese 1976a, 1976b). The founding members were mainly New York chemists, which led to discussions about how national the society should become. Prominent chemists were absent from the group of founders (Browne & Weeks 1952, 22).

The ACS' early years were characterized by struggle: many chemists saw it as purely a New York organization, leading to a decrease in membership during the 1880s. A rival emerged in the Chemical Society of Washington (1884), whereas the Chemical Section of the AAAS prospered between 1877 and 1887 (Beardsley 1964, 26). Within the AAAS Subsection for chemistry, attempts were made to organize local groups into a national society, in turn triggering much-needed reform and the reorganization of the ACS (Browne & Weeks 1952, 31; see also Bohning 1989, 1990, 2001). The ensuing success of the ACS as a national home for chemists is attributed to it adopting similar organizational structures to those of the British Society for Chemical Industry, characterized by local sections and national meetings at different venues, thereby covering the whole country and fostering communication between widely dispersed chemists (Saltzman 2006, 24; Beardsley 1964, 29; Browne & Weeks 1952, 81ff). Without a doubt, by the start of the twentieth century, the ACS had become the overarching organizational home for American chemists (Figure 4.1).

While other chemical societies sometimes proved dangerous for the ACS, control was further consolidated through specialization in divisions (Table 4.2). The creation of the first divisions in 1908 is generally seen as a reaction to the establishment of the American Electrochemical Society (AES) on the one hand, and the American Section of the Society of Chemical Industry (SoCI) and the American Institute of Chemical Engineers (AIChE; Thackray et al. 1985, 181) on the other hand. Concerns about industrial science and the industrial chemists, in particular, became a pressing matter in these years, with the ACS Division of Industrial and Engineering Chemistry being an attempt to unify academic and industrial chemists under one organizational roof, and to tend to the interests of industrial chemists who had not had any representation in the ACS before (Browne & Weeks 1952, 83). The first chairman of the division was Arthur D. Little, the famous consulting chemist. Looking at the further creation of ACS Divisions illustrates the changing makeup of the

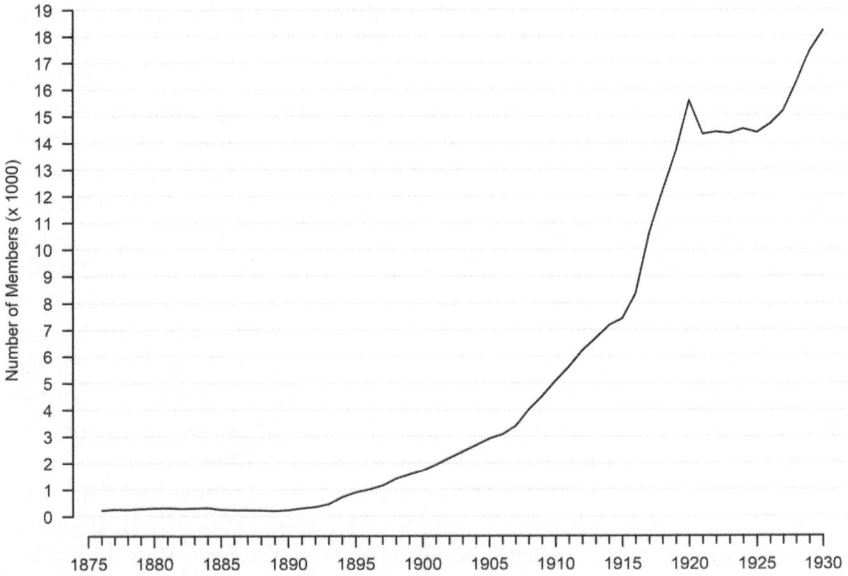

Figure 4.1 Annual Membership of the ACS, 1876–1930.
Source: Thackray et al. (1985, 249ff, table 2.4.)

Table 4.2 Specialty Divisions of the American Chemical Society, 1870–1930

Division	*Year established*
Industrial and Engineering Chemistry	1908
Agricultural and Food Chemistry	1908
Fertilizer and Soil Chemistry	1908
Organic Chemistry	1908
Physical Chemistry	1908
Medical Chemistry	1909
Rubber	1909
Biological Chemistry	1913
Environmental Chemistry	1913
Carbohydrate Chemistry	1919
Cellulose, Paper, and Textile	1919
Dye Chemistry	1919
Leather and Gelatin Chemistry	1919
Chemical Education	1921
History of Chemistry	1921
Fuel Chemistry	1922
Petroleum Chemistry	1922
Organic Coatings and Plastics Chemistry	1923
Colloid and Surface Chemistry	1926

Source: Thackray et al. (1985, 443ff, table 6.33).

American chemical profession, the advancement of chemical knowledge, and the complex interplay of scientific, political, and industrial interests in such an organization.

The next chemical society after the ACS was the Association of Agricultural Chemists (1984), its main goal being the creation of standards in methods and procedures for agricultural chemists (Bogert 1908, 174f). The outgoing nineteenth and early twentieth century saw an explosion of organizations, not all of them being academic societies strictly speaking, with some of them being occupational unions of chemists in various fields: the American Society for Testing Materials (1899), the American Ceramic Society (1899), the New England Association of Chemistry Teachers (1899), and the American Leather Chemists Society (1903). One of the more interesting organizations created at that time was the Chemists' Club (1898), a New York City-based society that set out to "provide both a sense of community and a collaborative forum for corporate executives, entrepreneurs, academics, and others concerned with serving and supporting the growth of science-related industries" (Records of the Chemists' Club, Board of Trustees Minutes, 1896–1914). Boasting a well-stocked library, explicitly wanting to serve the chemicals and allied industries (Records of the Chemists' Club, Board of Trustees Minutes, 1909–1916), and being used by both the ACS and the SoCI as a meeting place, the Chemists' Club soon counted prominent industrial chemists, such as GE's Whitney and Langmuir, also Edison, among its members.

As mentioned, the American Section of SoCI as well as the AES (founded 1902) were seen as competitors to the ACS. The AES brought together not only chemists but also engineers and metallurgists. At the time, electrical engineers were not granted full membership at the ACS (Bogert 1908, 177). The establishment of a specialized, dedicated society illustrates the growing importance of electrochemistry and the electrochemical industry (Trescott 1981).

Another very important organization on the field level that illustrates the changing environment and struggles for differentiation within chemistry is the AIChE, created in 1908. The Institute was born out of rivaling traditions: those of industrial and applied chemistry, and mechanical engineering (Reynolds 1983, 1). The newly developing discipline of chemical engineering led to tensions between chemical engineers, industrial chemists, and university chemists within the ACS, especially with regard to the relationship between chemistry and chemical engineering – some favoring chemistry, other emphasizing the engineering element. The chemical engineers struggled to establish themselves amid "lower" chemical analysts and mechanical engineers, and the chemical engineer's professional identity was one of the main reasons for the birth of the AIChE (Van Antwerpen & Fourdrinier 1958, 27; Furter 1980b). The founder of the Chemical Engineer Journal, Richard K. Meade, eventually chose to establish the Institute, despite opposition from the ACS, who reacted with the creation of their own Division of Industrial and Engineering Chemistry and its respective journal. In the early meetings, the pressing need arose to clarify how chemical engineering was to be defined, as well as the necessity to avoid conflict with the much more powerful ACS. The early AIChE members chose a strategy of very restrictive membership, expecting 5–10 years of industrial experience of its members, in order to not draw away membership from the ACS, as well as underline the Institute's role as complementary to the ACS (Reynolds 1983, 6). The strict membership provisions made the AIChE attractive to executives in the chemical industries, who were malcontent with the ACS's academic dominance (Van Antwerpen & Fourdrinier 1958, 31ff). The early years were characterized by slender resources and limited meetings,[1] but over time, chemical engineering separated somewhat from chemistry with its own body of scientific knowledge and the central intellectual concept of "unit operations" around, which the field was organized (Reynolds 1991b, 184; Van Antwerpen 1980). A strong focus on chemical engineering education was used to gain legitimacy as a subdiscipline of engineering (Van Antwerpen & Fourdrinier 1958, 50ff). By 1930, chemical engineering can be seen as an established and legitimate discipline that was perceived as a fifth classification of basic engineering (Reynolds 1983, 20), encompassing both chemistry and engineering, which was often at the heart of industrial chemical concerns and due to such an interdisciplinary position of special interests to this study.

The American Institute of Chemists (AIC) was organized in 1923 – originally envisioned as an economic organization of chemists, over time it shifted focus toward questions of professional ethics and conduct. This can be seen as an attempt to place the chemical profession upon a "real professional basis" similar to that of doctors or lawyers (Records of the AIC, file 1–14; November 28, 1928 letter from chairman to C. H. Strong). For admission to the Institute, both chemical training and work experience were expected (Thackray et al. 1985, 32; see also Carmichael 1974). Next to its professionalization-related activities, the AIC also eventually awarded the AIC Medal for chemists in governmental, academic, and industrial chemistry. After WWI, especially in the wake of the horrors of chemical warfare, the tendencies toward international cooperation grew stronger, leading to the formation of the International Union of Pure and Applied Chemistry (IUPAC) in 1919, and taking the place of previous organizational attempts at coordination beyond national borders. The IUPAC's aims were, among others, to coordinate scientific and technological activities of adhering member countries and their respective chemical associations, to set standards for symbols, notations, and constants, to provide documentation for industrial products, and to contribute to the advancement of chemistry in general (Fennell 1994, 35ff).

In summary, the field of US academic chemistry saw strong organizational growth from 1870 to 1930 that illustrated the specialization of the discipline, some of which was absorbed by the increasingly branched divisional structure of the ACS, while the creation of new divisions was often triggered by either scientific developments or rival organizations. But the ACS is not the only organization we need to keep in mind when assessing the field since chemistry at the beginning of the twentieth century had become too complex in its dispersal throughout society to be adequately covered by an academic society. Furthermore, the interrelations of chemistry with engineering and metallurgy led to different groups pursuing different aims, as well as contested professional identities and lack of representation within the ACS often causing new organizations to be born. Thus, the first layer in this timeline of chemistry offers valuable clues about relevant actors and interests within the field, especially concerning the growing importance of the industrial chemists and their corporate advocates.

Chemical Journals

To further assess the development, and especially differentiation, of American chemistry, one needs to account for scientific publications as indicators of not only the expansion of knowledge but also the growth of the discipline in general. Books – be they monographs or edited volumes – certainly played a role in American chemistry as well, but they will not be included in this analysis. Evidence from reference works taking stock of chemical publications (Lovelace & Thomas 1919, 1920; Crane et al. 1927; Bolton [1899] 1967a, [1903] 1967b, [1893] 2005) suggests a growth in the amount and intensity of chemistry-related books being published, yet no quantitative analysis of the time period in question exists to my knowledge. At this point, we shall investigate how scientific journals and related periodicals were established since they are commonly regarded as the central avenue of scientific communication (Fyfe 2016). The history of American chemical journals follows a similar trajectory to that of scientific societies: one can perceive a trend from the personal to the organizational, that is, early journals were highly dependent on the founder/publisher and often ceased publication once funds lacked or the founder died, whereas the late nineteenth century saw an increasing number of journals established that served as organs of organizations, giving them a more stable footing (Table 4.3).

Chemists who wanted to publish their findings before 1870 had a variety of options. They could publish in journals such as the American Journal of Science (AJS), or they could choose the proceedings of the national science organizations. Furthermore, in the 1860s new journals, for example, the Journal of the Franklin Institute, entered the field (Beardsley 1964, 35). Yet many chose to look past the United States and publish abroad, mainly in Germany, for the lack of dedicated American chemical publications at the time (Saltzman 2006, 23;

Table 4.3 US Chemical Journals, 1870–1930

Journal	*First year of publication*	*Ceased publication/ still active in 1930*
Journal of Applied Chemistry	1866	1875
American Chemist	1870	1876
The Laboratory	1874	1876
Proceedings of the ACS	1876	–
American Chemical Journal	1879	1914
Journal of the ACS	1879	–
Journal of SoCI (UK)	1882	–
AOAC Methods	1884	–
Bulletin of the Chemical Society of Washington	1884	1895
American Analyst	1885	1892
Journal of Analytical Chemistry	1887	1893
Journal of Physical Chemistry	1887	–
Electrochemical Industry	1902	–
Transactions of the American Electrochemical Society	1902	–
Chemical Engineer	1904	–
Journal of Biological Chemistry	1906	–
Chemical Abstracts	1906	–
Journal of the American Leather Chemists Association	1906	–
AIChE Transactions	1908	–
Journal of Engineering and Industrial Chemistry	1908	–
Chemist-Analyst	1911	–
Journal of AOAC	1915	–
Chemical & Engineering News	1923	–
The Chemist	1923	–
Journal of Chemical Education	1924	–
Analytical Chemistry	1929	–

Sources: Bogert (1908), Crane et al. (1927), Beardsley (1964, 34ff), Bolton ([1899] 1967a, [1903] 1967b, [1893] 2005), Ihde (1964, 270ff), Thackray et al. (1985, 176ff).

Thackray et al. 1985, 176). In Germany, a variety of journals existed that tended to different specificities of chemistry: Annalen der Chemie, Journal für praktische Chemie, Zeitschrift für Chemie, and the Chemische Berichte, to name the most relevant ones. The total amount of publications in foreign journals tripled between the time periods of 1879–1891 and 1892–1914. This effect must be seen as a consequence of both the differing lengths of the time periods and the strong growth of American chemical publications. As Crane's analysis of statistics based on Chemical Abstracts shows, chemical publications took off with a steep trajectory, curtailed only temporarily by the war. Throughout WWI, American shares of chemical publications grew steeply, further boosted by the German defeat, although these effects tapered off during the 1920s. Yet Germany was not able to reclaim its dominant position, with the United States and Germany now settling at about level in publications. The absolute numbers show a powerful increase in articles abstracted (Crane 1944; see also Thackray et al. 1985, 401f, table 6.8).

The Boston-based Journal of Applied Chemistry was one of if, not the first dedicated journal for chemistry, established in 1866 and ceasing publication in 1876 (Thackray et al. 1985, 177), yet there is hardly any evidence of its distribution reaching further than the confines of the city of Boston. Lacking national and, most importantly, stable organizations that could serve as centers for communication and to suppliers of the necessary funds for printing and distributing periodicals, the landscape for chemical publications remained chaotic until 1870, when the situation changed and it had become clear that general scientific journals would no longer suffice for the increasingly specialized interests of the chemists. The American Chemist was established as a supplement to the American reprint edition of the British Chemical News by the Chandler brothers (Thackray et al. 1985, 177; Beardsley 1964, 36; Crane et al. 1927, 69). While the journal encouraged the professional interests of the chemist, it failed to establish relations with industry, and financial losses led to its end seven years later, in 1877 (Beardsley 1964, 37). With the birth of the ACS, the Proceedings of the ACS started publication in 1876, and only three years later came its main organ: the Journal of the ACS (JACS, established in 1879). Lacking contributions in its early years, the Journal experienced a similarly bumpy start as the ACS, which discouraged many chemists from publishing there (Beardsley 1964, 39; Kevles 1979, 143).

1879 also saw the first edition of the American Chemical Journal, established by the eminent chemist Ira Remsen, who used the journal as an outlet for his own and other chemical research at Johns Hopkins (Saltzman 2006, 24). His journal "outclassed" (Beardsley 1964, 40; see also Crane et al. 1927, 69) JACS in the 1880s, existing in parallel for three decades until 1914 when the two journals merged. Indeed, it was another merger that helped establish JACS as the strong, central voice of American chemistry: the Journal of Analytical Chemistry, established in 1887, merged with JACS in 1893 and boasted JACS' subscriber base considerably (Thackray et al. 1985, 177). The late nineteenth and early twentieth century saw a differentiation of journals, similar to that evidenced by chemical societies. Furthermore, newly founded societies and professional organizations of chemists created their own organs, such as the Transactions of the American Electrochemical Society (1902). Journals occupying a position somewhere in between science and industry, or catering to the special interests of the chemical engineer, found their audiences after 1900: Electrochemical Industry (1902), Chemical Engineer (1904), and Chemist-Analyst (1911); with chemical engineers adding another publication with the foundation of AIChE (AIChE Transactions, 1908). The struggle for professional identities, especially by the industrial chemist, as outlined previously, that leads to the founding of the ACS Division of Industrial and Engineering Chemistry, also brought with it a dedicated journal: the Journal of Engineering and Industrial Chemistry (1908). As a reaction to the explosion of chemical research and publications, the ACS started its abstracting service, Chemical Abstracts, in 1906, as a means to keep abreast of the developments in the discipline (Thackray et al. 1985, 180). In the 1920s, the ACS added its own "news" service, with Chemical & Engineering News (1923), a weekly news magazine (Thackray et al. 1985, 181). In summary, similar to the differentiation seen on the level of scientific and professional societies, the available platforms for periodical publication in chemistry exploded with the turn of the century. A variety of journals were created not only pertaining to different specialties of chemistry but also to industry and different professional groups of chemists. Concurrently, the amount of chemical publications grew by several magnitudes, and the United States became a major player in chemistry.

Chemistry in Higher Education

To conclude this assessment of the academic/scientific layer of the field of chemistry in the United States, we shall take a look at chemistry graduates and the centers of American chemistry.

Taking stock of graduates is a good way to illustrate the growth of the discipline, and to obtain an understanding of the way that academically trained chemists flooded into the industry, an argument sometimes found in explanations of the birth of the industrial research laboratory, as outlined previously. The numbers regarding graduates with chemical degrees, be they a bachelor, a second-level degree, or a PhD all show a similar trend between 1890 and 1930: absolute growth but relative decline (Figures 4.2 and 4.3). Thackray et al. find exponential growth on all levels for all subjects during this time period, not just for chemistry (Thackray et al. 1985, 40; see also Weart 1979). Population growth is part of the reason, but it is outpaced by degrees: When calculated as a unit of population, chemistry degrees quadruple, from 1200 degrees per million of the population aged 20–24 in 1890, to 4800 in 1930 (Thackray et al. 1985, 45; see also tables 3.4, 3.5). The decline of chemistry degrees as a percentage of all degrees indicates that, while the number of graduates with a chemistry degree increased, indeed creating a pool of manpower that was available for positions in industry, the relative importance of chemistry at American universities declined.

Next, some data are available regarding not only a population level but also an organizational level, namely doctorates awarded by different

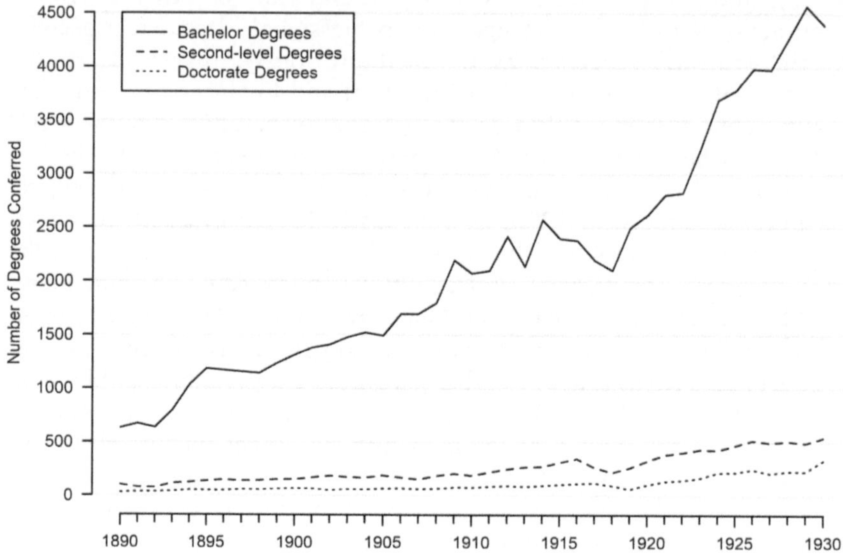

Figure 4.2 Bachelors, Second-Level Degrees, PhDs in Chemistry, 1890–1930.
Source: Thackray et al. (1985, 257ff, tables 3.1, 3.2, 3.3); Degree Definition (see Thackray et al. 1985, 40n4).

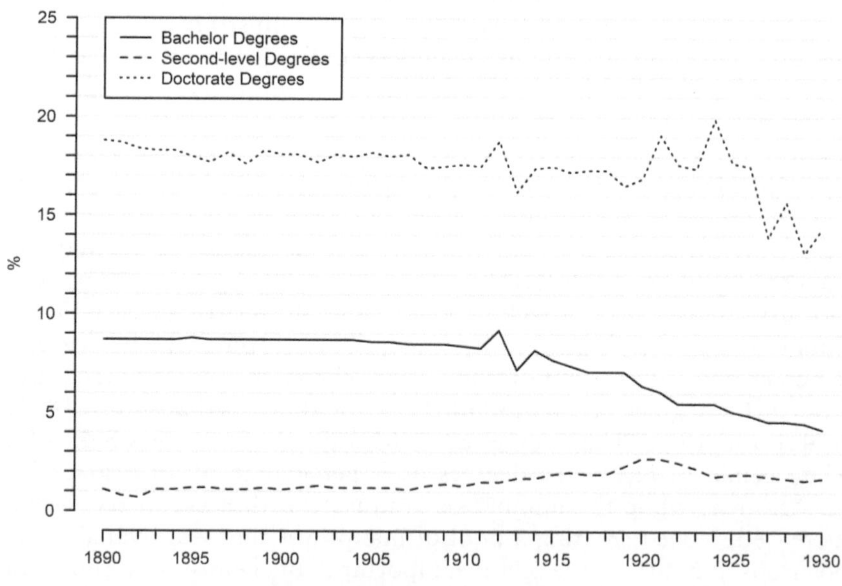

Figure 4.3 Bachelors, Second-Level Degrees, PhDs in Chemistry as Percentage of all Degree Conferrals, 1890–1930.
Source: Thackray et al. (1985, 257ff, tables 3.1, 3.2, 3.3).

universities over the years. Thackray et al. (1985, 147ff; see also table 6.2) cover the period from 1861 to 1899, whereas Kevles (1979, table 9) compares the years prior to 1897 with those between 1898 and 1915, covering the most important part of the period of interest for this study, and right up until WWI. A rough picture of the landscape of American higher education, and academic research in chemistry in particular, emerges, especially once compared to the steep decline of doctorates awarded abroad. The common practice of doing PhD-level research in Germany ended with the gradual build-up of capacities for teaching and research in the United States. Chemical faculty increased concurrently with the growth in graduates and universities awarding doctorates as well. Reliable estimates are hard to come by or gauge, mostly due to the vagueness of the categories of "faculty" or "college teacher" (Thackray et al. 1985, 141), yet some inferences can be drawn, mostly based on ACS surveys. In absolute terms, chemistry faculty grew considerably in the 60 years between 1870 and 1930, while remaining mostly stable in the years before and after 1900 when calculated as a percentage of the total faculty, and rising after WWI (ibid., 141f, figures 5.3-1, 5.3-2). The number of chemists appointed to college presidencies reached a peak during the 1910s and remained strong throughout the 1920s, surely a result of WWI and chemical boosterism (ibid., 153, figure 6.1-4).

Chemistry in Industry: Companies, Products, and Processes

Sketching the growth and development of the American chemical industry is a harrowing task that requires a clear definition of what "chemical industry" means in the first place, which is not easily arrived at:

> Unlike industries which sell a limited variety of products (e.g., automobiles or cigarettes) or serve common markets (like household appliances or medical supplies), the chemical industry is an amalgam of highly diverse enterprises, manufacturing an astonishing array of intermediates and final products for a multitude of markets ....
>
> (Thackray et al. 1985, 84)

Two general terms often found divide the chemical industry into "chemical process industries" on the one hand, and "chemicals and allied products" on the other (ibid.). The former designates a group of industries that use chemical processes for creating their goods, such as the paper, petroleum, and rubber industries, whereas the latter denotes a very diverse group of industries that produce chemical products, "ranging from pesticides to perfumes" (ibid.). The most important developments in industrial chemistry from 1870 to 1930 can be traced along these lines. Two of the main transformation of the chemical industry as a whole fall on the side of processes: First, processes for the artificial production of natural chemical substances were developed, for example, making the production of synthetic indigo possible. Second, existing processes were made more efficient – the production of soda ash was revolutionized by the Solvay Process, to name one example. Regarding products, the main change in this industry surely was the ability to create artificial chemicals not found in nature, such as synthetic dyes (Murmann 2002, 401f; Aftalion 1991, 32ff). "Chemicals and allied products" have been covered by the United States Census since 1890, and following Thackray et al., along with the availability of the data, the discussion of the quantitative developments of chemistry utilization in industry shall mainly focus on chemicals and allied products and be extended to the chemical process industries where information is on hand – with the usual caveats applying to the historical data.

The Development of the Chemical Industry

Between 1870 and 1930, the number of establishments in the American chemical industry grew steadily, a marked uptick especially notable after WWI, when the number of firms almost doubled between 1914 and 1919 (Figure 4.4). Even though direct comparisons of the early decades with those post-1900 are to be taken with care,[2] a general trend of continuous growth can be perceived.

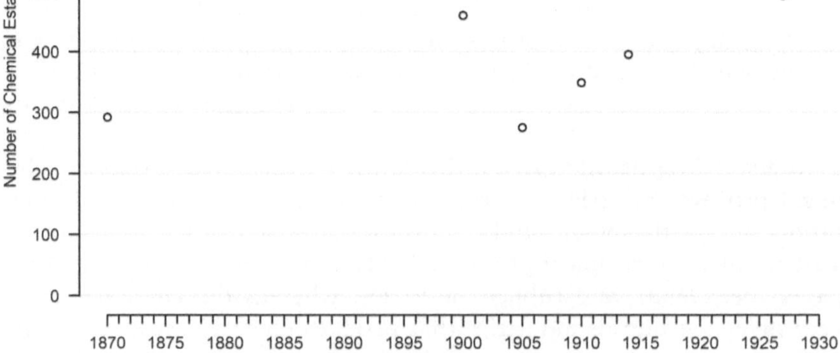

Figure 4.4 Number of Chemical Establishments, 1870–1930.
Sources: Haynes (1954a, 401; 1954b, 277; 1954c, 450).

The chemical outputs of the entire industry grew uninterrupted from 1899 to 1930, with a similar uptick of pace visible after WWI corresponding to the number of chemical firms, with chemical outputs eventually overtaking total manufacturing at the beginning of the 1930s (Figure 4.5).

Of course, this rapid growth needs to be seen against the backdrop of the fundamental transformation of American industry and society outlined previously (Haynes 1954a, 244). Looking at capital investment to gauge the relative growth of the chemical industries, the chemicals and allied product industries increased their share from 4.3% in 1879 to 6.7% by 1929 as the percentage of capital invested in total manufacturing (Figure 4.6). Similarly, although less dramatically, the chemical process industries increased their share from 41% to 55%. While in the early years, food and primary metals accounted for most of the capital investment in the chemical process industries, by 1929 chemicals and allied products, and petroleum had grown considerably (Thackray et al. 1985, 89f).

With the start of the twentieth century, the chemicals and allied products industry had grown to be increasingly controlled by oligopolies, with the product value that was controlled by these oligopolies almost doubling between 1909 and 1927. Both the chemical process industries and chemicals and allied products made up a considerable amount of the

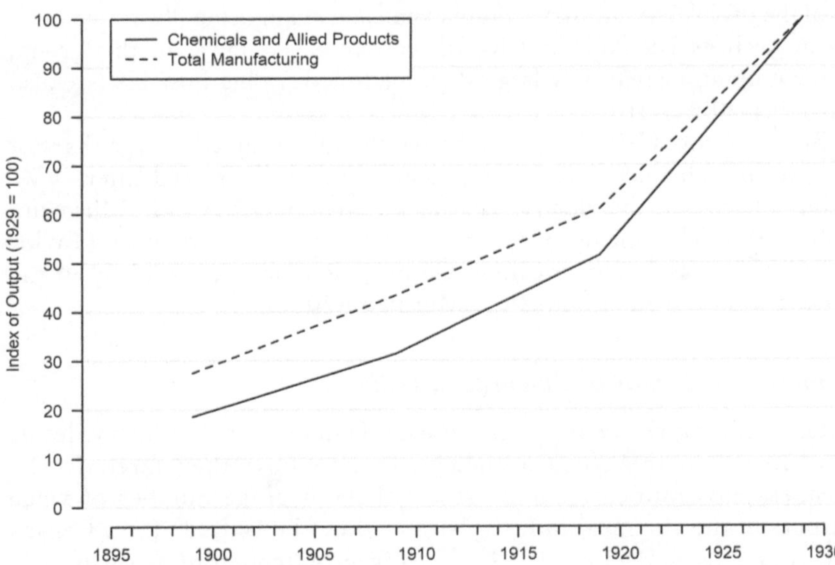

Figure 4.5 Outputs of Chemical and All Manufacturing Industries, 1899–1929 (Base Year: 1929).

Source: Thackray et al. (1985, 313, table 4.1).

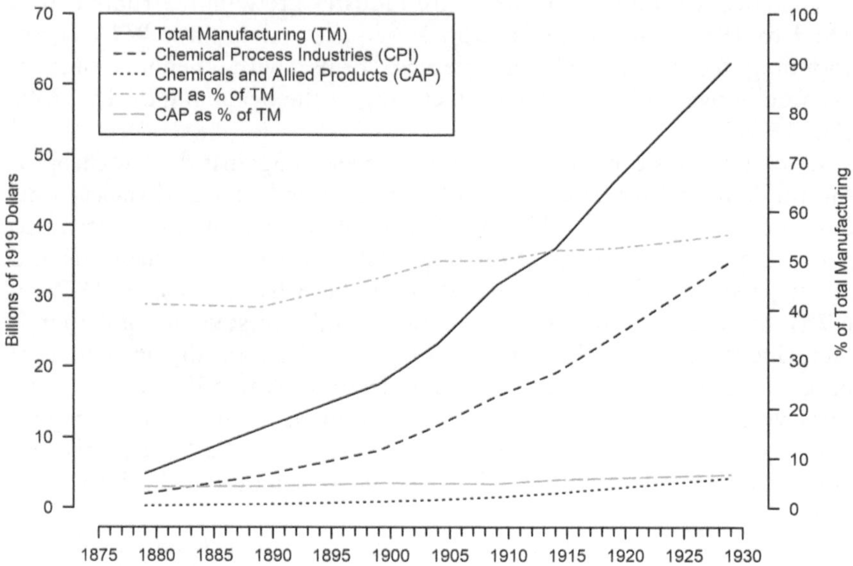

Figure 4.6 Capital Investment in the Chemical Industries and in Total Manufacturing, 1879–1929.
Source: Thackray et al. (1985, 314, table 4.2).

top 100 industrial corporations (Thackray et al. 1985, 90, 319, table 4.6), with the petroleum industry's rapid expansion again being illustrated here. Firms such as Du Pont and Kodak continuously stayed in this group, making up an increasingly large share of American big business (see also Chandler 2005, 41ff).

By the eve of WWI, the United States already occupied a large share of the world's chemical production, which further increased after WWI being fueled by the demise of the German industry (cf. Murmann 2002, 400). The seizing of German patents after the armistice (Taylor 1976, 273) was key in breaking German dominance, especially in the production of synthetic dyes (cf. Murmann 2002).

Employment Trends of Chemists in Industry

After outlining the changes in the American chemical industry, let us now focus on the chemist themselves and how they fared in the business environment. Taking stock of the average number of wage earners in the chemical industry supplied by the United States Census (Haynes 1954a, 1954b, 1954c), steady growth before WWI is supplanted by a steep jump that more than doubled the number of chemical employees by 1919, followed by a decline in the immediate aftermath of the war.

How many of those wage earners in industry were actual, academically trained chemists, and how many were shop workers or administrative personnel, is unfortunately harder to assess. For this purpose, Figure 4.7 also shows the Census estimates for the number of chemists between 1870 and 1930, which once again was subject to changing classifications with new disciplines emerging such as chemical engineering, and the grouping of chemists with metallurgists and assayers (Thackray et al. 1985, 13ff, 204ff). The lower numbers of Census Chemists compared to wage earners in the chemical industries indicate the reliance of industry on non-chemically trained personnel, especially during its early years and later during WWI when the wartime demands made an influx of untrained, temporary personnel likely. Contrarily to wage earners in the chemical industry, the rising trajectory of chemists did not see a drop after WWI, which suggests that while industry scaled back after the intensity of the war effort, chemists still found homes in academia and the emerging field of federal support for science and specifically chemistry.

As the discussion of research personnel in industrial research laboratories earlier has shown, chemists made up the largest share of industrial researchers from the earliest NRC surveys onward. Their ranks grew in absolute numbers, but compared to engineers the chemists lost some of

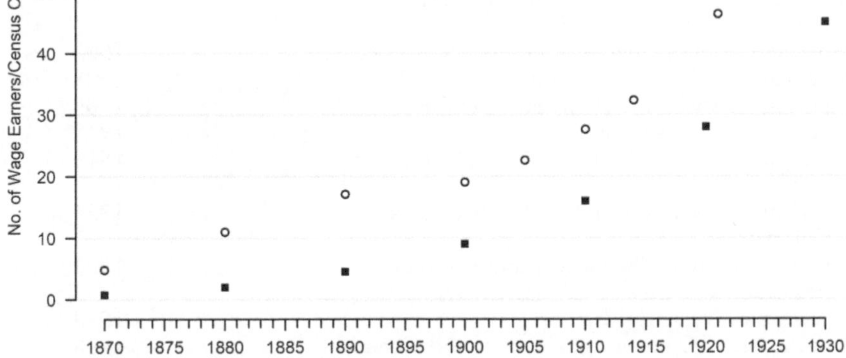

Figure 4.7 Average Number of Wage Earners & Census Chemists, 1870–1930. Sources: Haynes (1954a, 401; 1954b, 277; 1954c, 450), Thackray et al. (1985, 247, table 2.2).

their share of total research personnel, from one in three in 1921 to roughly one in four in 1931, yet they still maintained their central position (Thackray et al. 1985, 120).

Trade Associations and Other Organizations in the Chemical Industry

Of course, professional associations and other organizations shaped the field of chemical industry during the period of interest. The Manufacturing Chemists' Association (MCA) is America's oldest trade association. Founded in 1872 to protect chemical manufacturers against problematic legislation and unjust freight discrimination, as well as to promote and aid the general interests of the chemical industry (Manufacturing Chemists' Association Inc. 1972, 7), its ranks soon swelled to 41 members by 1908 (Bogert 1908, 168; Table 4.4).

In subsequent years, the MCA not only conducted scientific and technical studies but also lobbied congress in tariff matters (Manufacturing Chemists' Association Inc. 1972, 30ff). Its foundation can be seen as the first signs of organized action by chemical industry and the establishment of a field of industrial chemistry in the nineteenth century (Haynes 1954a, 251). As indicated earlier, industrial chemistry and the interests of chemical manufacturers were not of strong concern in the early years of the ACS, which lead to the creation of the American

Table 4.4 Foundations of Trade Associations and Other Professional Organizations, 1870–1930

Organization	*Year established*
Manufacturing Chemists' Association	1872
American Section of Society of Chemical Industry (UK)	1894
American Leather Chemists Society	1903
American Oil Chemists Society	1908
American Association of Cereal Chemists	1909
American Dyes Institute	1918
Société de Chimie Industrielle (FR) (American Section)	1919
American Association of Textile Chemists and Colorists	1921
Synthetic Organic Chemical Manufacturers' Association	1921
Directors of Industrial Research	1923
Association of Consulting Chemists and Chemical Engineers Inc.	1928

Sources: Bogert (1908), Thackray et al. (1985), Haynes (1954a, 1954b, 1954c), own research.

Section of SoCI in 1894, which in turn sparked reactions by the ACS to further the inclusion of industrial chemistry in their own ranks with the Section of Industrial and Engineering Chemistry. The birth of SoCI in America was especially fueled by a perceived discrimination by the ACS against articles submitted for publication by industrial chemists, as well as the admittance of manufacturers only as associates and not full members (Bogert 1908, 175). Over the years, a plethora of professional organizations was established, some already mentioned previously, yet some of them became important players in the industrial field by representing their particular branches of industry. In particular, the American Dyes Institute, with Du Pont as a leading member, was very active in lobbying Congress to enact an embargo on synthetic dye imports, which shaped the industry in the post-war years (Rhees 1987, 265ff; Steen 2014).

One newly formed organization that was of special interest not only to the business of chemistry but also to the business of chemical research in industry is the Directors of Industrial Research (DIR) group. The idea appears to have been conceived when the Nela Park Laboratory's director visited Du Pont, with the DIR being officially founded in 1923 upon invitation from Robert Yerkes of Yale and the NRC, as well as Reese of Du Pont and Alfred D. Flinn of the Engineering Foundation and the NRC (David A. Hounshell and John K. Smith Research Notes for Science and Corporate Strategy (Accession 1850), box 2). Many of the early DIR members had gotten used to exchanging ideas through the war effort and wanted to continue the fertile association and sharing of information (cf. Hoddeson 1977, 26). The founding members were, among others, the corporations Western Electric (AT&T's engineering arm), Kodak, Du Pont, Westinghouse, and GE (ibid.). Committed to scientific research, the DIR had the interests of the application of research to the aims and needs of industry at heart, and was subsequently concerned with the question of how best to organize research in an industrial laboratory setting. The founding of the DIR not only extends our understanding of the field that chemical corporations found themselves in between 1870 and 1930 but also serves as a clue to how the diffusion of the industrial research laboratory transformed the organizational field. After WWI, the laboratory had become an expected element of chemical corporations of a certain size, with the existence of a variety of laboratories now requiring new means of communication and exchanges of expertise as well as giving birth to special organizations that furthered a research-focused agenda within the chemical industry, such as the DIR. As a conjecture, the leadership of the four pioneers may also have been amplified by such organizations, since the archetypal research directors, such as Whitney and Mees, could spread their vision of science in a corporate setting.

Chemistry and Politics: WWI, Boosterism, and the New Landscape of Federal Support for Science

Before diving into the relations between chemistry and politics – focusing first on the place of the chemist in governmental employ and the organizations supporting them, and second on the effects of WWI – a brief quantitative look at how chemists fared in the federal government is in order. As before, reliable data are hard to come by for the early years. The chiefly responsible factor is the large variety of classification schemes used by federal departments prior to 1923. Thackray et al. chose a "pragmatic criterion" (Thackray et al. 1985, 26) to deal with these idiosyncrasies: They relied upon the government's definitions of "chemist" as a distinct bureaucratic category (ibid.). While some generalizations are possible based on these data, precautions are necessary when trying to compare these figures to the Census numbers.

As can clearly be seen (Figure 4.8), the number of chemists on a federal level only grew slowly leading up to and including the early decades of the twentieth century. With the adoption of "many novel regulatory and service functions" (Thackray et al. 1985, 130) by the federal government to cope with the economic and societal transformations since the Civil War, more chemists were needed as well. The years leading up to WWI and the United States' entering of the war in 1917 proved to be catalysts for a stark increase in the number of chemists employed on the federal

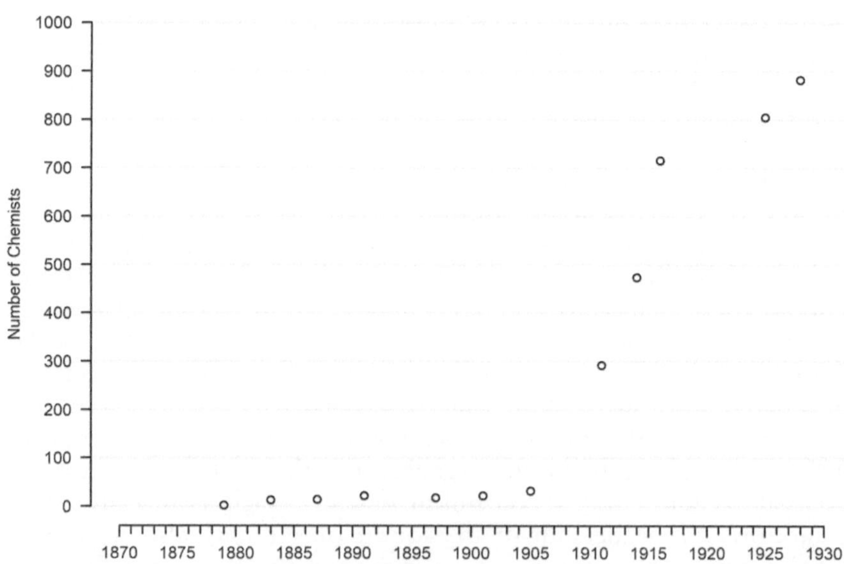

Figure 4.8 Chemists in Federal Government, 1879–1930.
Source: Thackray et al. (1985, 358ff, table 5.13).

level. In particular, the creation of an entirely new organizational landscape of science funding and support for science that was brought about in the wake of WWI will be discussed later as major factors not only in rapidly increasing the number of chemists in government but also in securing a novel position for chemists and chemistry in national politics that was sustained in the interwar years.

Chemists in Federal Employment

As Beardsley (1964, 55) notes, the relationship between chemists and the (federal) government before the twentieth century can roughly be divided into three phases. In the first phase, before the Civil War, most chemists who did not pursue academic careers would find employment at the United States Geological Survey and, to a much lesser extent, at agricultural boards and assaying offices. These chemists were expected to be skilled in metallurgy and geology at a time when the specialization of sciences had not proceeded very far (ibid., 49ff). Apart from the surveys, few notable exceptions existed, such as the Smithsonian Chemical Laboratory (ibid., 55; see also Dupree 1957, 91ff; Pursell 1966; Kevles 1995, 10ff).

The second phase began after the Civil War. With the passage of the Morrill Act of 1862, the Department of Agriculture (USDA) was founded, which came to be one of the main employers of chemists in the following decades (Dupree 1957, 149ff). But the Morrill Act also laid the foundations for many agricultural experiment stations in a variety of states, which sought out the skills of the chemists and over the year slowly transformed into chemical research centers (cf. Rosenberg 1976, 135ff). With the 1880s, the third phase came and was designated by a more stable organization of the federal relationship with science. New surveys such as the Coast and Geodetic Survey were conducted, the Weather Service was founded, and the Naval Observatory rose to further prominence (Kevles 2013, 15; Pursell 1966, 224ff). The agricultural experiment stations were further strengthened by the Hatch Act of 1887. The Geological Survey established chemical laboratories in San Francisco, Denver, and Washington, supporting not only routine work but also chemical research (Taylor 1976, 276). According to Beardsley, it was in this phase that the services of the chemists found increasing appreciation, as well as permanence in positions (Beardsley 1964, 57).

With the beginning of the twentieth century, the demand for chemists intensified (Table 4.5). In 1901, the Bureau of Chemistry was founded as a department of the USDA – its aims being the establishment of standards, as well as the employing of chemical theory and methods to the testing of foods and drugs. This Bureau was instrumental in the passage of the Pure Food and Drug Act in 1906 and was subsequently tasked with the inspection of products via chemical analysis which had grown into a much greater role on the national level. By 1908, 425 chemists

Table 4.5 Foundations of Governmental Organizations Related to Chemistry, 1870–1930

Organization	*Year established*
U.S. Department of Agriculture	1862
National Bureau of Standards	1901
Bureau of Mines	1910
Naval Consulting Board	1915
National Research Council	1916
The Chemical Foundation	1919
Science Service	1920
National Research Endowment (National Research Fund)	1925

Sources: Beardsley (1964), Thackray et al. (1985), own research.

were employed in the Bureau (Taylor 1976, 276). Up to and during WWI, the USDA maintained its centrality as the employer of chemists within the federal government, even though the share of chemists employed declined relative to other government agencies, especially because of the rapid expansion of the Department of War and the Navy during and after WWI (Thackray et al. 1985, 131, 132, figure 5.2-3; Rae 1979, 257). Further opportunities for chemists arose with the creation of new governmental organizations, such as the Bureau of Standards. Founded in 1901, it was tasked with the testing of steel and structural materials, as well as the promotion of accurate methods of analysis (Thackray et al. 1985, 130; Pursell 1966, 233; Dupree 1957, 271ff). Seeking new research functions in the prewar years, it acted as a "sort of bridge between government and industry" (Taylor 1976, 276). Oversight of the activities of the Bureau of Standards, the Patent Office, the Bureau of Mines, and some other agencies was moved to Herbert Hoover's Department of Commerce in the 1920s, who sought to promote industrial research and incentivize research projects as well as cooperation within industry (Dupree 1957, 336ff). Of course, chemists were not only active at a federal level but also at the state and local governmental levels that were also offering jobs, such as with the Boards of Health. Unfortunately, prior to 1950, data on the quantity of chemists engaged at this level are not available (Thackray et al. 1985, 135).

As can be seen, the pace and intensity of federal support for science increased after the Civil War, spawning a wide number of bureaus and agencies. Many of these, especially those established prior to the Civil War, continued to exist with altered missions when their original purposes lost importance, leading to a "bewildering multiplicity" (Dupree 1957, 289) of governmental agencies by the time WWI started. As federal support for science increased, a variety of governmental agencies hired chemists in one way or another, often tasking them with routine functions and making use of their skills in testing and analysis. Yet the biggest shakeup came with WWI not only when the federal government began drawing heavily on the fruits of science for the war effort but also when chemists began leveraging their positions for increased political power and support of their own agendas.

The Effects of WWI

Without question, the Great War served as an important watershed moment for chemistry (cf. Kevles 1972, 7ff). The chemists' role in it was so eminent that in the aftermath and up until today WWI is often touted as being the "chemists' war" (Rhees 1993a, 41; LaFollette 1990, 9). Through the experiences of war, new areas of contact were created between politics, science, and industry, not only restructuring the organizational field but also modifying the perceptions and understandings of each other (cf. Pursell 1966, 236f). Following Rhees (1993a), the effects of WWI, while in reality complex and multilayered, can be analytically divided into five areas: industrialization, militarization, nationalization, politicization, and popularization.

Industrialization

Gearing up for war caused a strong increase in industrial activity that eventually reached unparalleled levels (Thackray et al. 1985, 99). In particular, the increased demand for explosives, munitions, and other chemicals was notable, as well as trained chemists available for work in plants and factories. Embargoes on German chemicals led to severe shortages and to the so-called "chemical famine" (Haynes 1954b, 36) (sometimes also called the "dye famine", cf. Rhees 1993a, 41) that was due to an over-dependence on raw and intermediate materials from abroad, and specifically from Germany (Steen 2014, 30). Thus, the American chemical industry transitioned from producing finished products to also incorporating raw materials into their production, a capital and knowledge-intensive process. In order to further coordination and mutual action by chemical manufacturers, the Chemical Alliance was formed in 1916 (Haynes 1954b, 50ff). Legislative efforts were made to protect and Americanize the chemical industries: The Trading with the Enemy Act of 1917 restricted trade with Germany and its allies, while the end of WWI brought with it the dye tariff campaign (Rhees 1987, 263ff; Steen 2014, 172ff), pushed for by a coalition of chemical manufacturers and other organizations, which sought a tariff to protect the newly built industry against re-awakening German competition, culminating in the passage of the Fordney-McCumber Tariff Act (1922). Indeed, with the armistice, the transformations of the chemical industry had not yet ended, as the dye tariff campaign illustrates. American industry was further bolstered by the seizing of several thousand German chemical patents by the Alien Property Custodian. The seized patents were sold – for a nominal fee – to an organization founded for just this specific purpose: The Chemical Foundation, Inc. (1919). Acting as a hybrid organization that combined the tasks of a government agency, trade association, and private foundation, it leased the seized patents to American firms (Rhees 1987, 272; Taylor 1976, 273). This competitive advantage allowed the American chemical industries to break free of the commanding position of the German chemical manufacturers. The Chemical Foundation spent royalties on former German patents to lobby for additional protection that was granted by Congress (Steen 2014, 203). Thus, WWI's effect on the American chemical industry was a strong growth in capacities, firms, and products, the creation of strong legislative protection, as well as the foundation of organizations aimed specifically at aiding and/or representing the industries and eventually leading to the United States emerging as the leading nation in industrial chemistry (Taylor 1976, 273).

Militarization

The militarization of American chemistry can be traced in two ways: corporations doing research and development for the military, and the foundation of new organizations building bridges between the military and academic/industrial chemists. All of the research pioneers, as well as other firms, were engaged to support the military during the war via research and development, such as AT&T's work on radio and other signaling equipment, and Du Pont's involvement in the development and production of poison gas – with the use of chemical warfare cementing itself in people's minds, thereby testifying to the destructive power of science, and furthering the perception of chemists being instrumental to the war effort.

Chemical warfare research was spread over a couple of military and civilian agencies, such as the Bureau of Mines, the Ordnance Department, the Medical Corps, the Signal Corps, and others (Rhees 1987, 168; see also Dupree 1957, 302ff). Furthermore, several military organizations were founded to coordinate activities, such as the Naval Consulting Board (NCB). Organized in 1915 by Josephus Daniels, Secretary of the Navy, at the proposal of Edison, the NCB brought together a variety of members of scientific and engineering societies to provide advice to the Navy and gear the nation toward war. The ACS chose GE's Whitney as their representative on the Board (Birr 1957, 63ff; Wise 1985, 169ff). The NCB was the chief promoter of the industry preparedness campaign prior to the United States entering the war. By the end of the war, one third of all American chemists were serving, or had served, in some federal agency, and 5400 chemists served in the Armed Forces by the end of 1918 (Rhees 1993a, 41f). During demobilization, the ACS even lobbied to keep the Warfare Service alive, since chemists had learned to appreciate the benefits of military patronage (ibid., 42; Rhees 1987, 199). Bringing the power of science, and specifically chemistry, to bear in the war led to academic and industrial chemists working in close cooperation, and encouraging "an unprecedented degree of unity and cooperation in the American chemical profession which continued into the postwar era, helping create a new sense of disciplinary identity and solidarity" (Rhees 1993a, 42). Furthermore, important relationships between academic/industrial chemistry and agencies of the Armed Forces were built, and proponents of industrial research, such as Whitney, were placed in key positions that amplified their reach and visibility (Rhees 1987, 175ff).[3]

Nationalization

Following Rhees (1993a), the American chemical community grew increasingly nationalized by the late 1910s. While this development is less important when accounting for the reshaping of the organizational landscape of chemistry up through WWI, any analysis scrutinizing discursive positions needs to be aware of nationalistic tendencies in chemistry from the outbreak of WWI onward. On the one hand, these tendencies manifested as chemists starting to account for their own history through the ACS Division for the History of Chemistry, and the ascendance of Priestly to the "patron saint" of American chemistry (Rhees 1993a, 44). On the other hand, anti-German acts such as the Chemists' Club prohibiting the use of the German language were testament to such currents of thought (ibid.; Records of the Chemists' Club, Board of Trustees Minutes, 1915–1929).

Politicization

The politicization of chemistry is meant as a general descriptor of the activities undertaken by chemists in industry and academia to influence policy and legislation, thus placing themselves in an important position within or vis-à-vis the federal government. During WWI this happened in a variety of ways, transforming chemistry as a profession both in academia and industry. As Rhees notes, the war was the first time the ACS – and other chemical societies and organizations – became involved in "matters of national policy" (Rhees 1993a, 42f). The main goal was not to secure government funding for research, but rather to ensure the protection of the chemical industries, which in turn were seen as a provider of jobs and funds for chemists (ibid., 43). It is from this perspective that the dye tariff campaign and other lobbying efforts need to be understood.

But there were also other pivotal developments brought about by the politicization of chemistry, namely the creation of an institutional framework that would administer support for science serve to mobilize the power of science (Dennis 1987, 507ff), and act in relation to the federal government, such as the NRC or the National Research Endowment. The NRC was organized at the behest of the astrophysicist George Ellery Hale who urged his fellow members to offer the Academy's services to President Wilson, who accepted the offer in 1916, establishing the NRC as an agency of the NAS (Kevles 1968, 431; Cochrane 1978, 209ff). The NRC's mission was to "undertake a national inventory of available scientific equipment and men, establish special committees to survey important problems for research, and promote cooperation between investigators in government bureaus, universities, research institutions, and industrial laboratories" (Cochrane 1978, 211). The NRC was essentially a federation of representatives from scientific societies, industrial and government laboratories, and universities (Kevles 1971, 185), and its Executive Committee was staffed by leading academic and industrial figures, including AT&T's Carty (Cochrane 1978, 214f). Next to the Chemistry Division, which was one of the "most energetic and effective sections" (Haynes 1954b, 40), was also the Committee for the Promotion of Industrial Research, headed by Carty. Preceding the end of the war, the NRC was put on a permanent basis by executive order of President Wilson (Kevles 1968, 432ff; Cochrane 1978, 233ff), with its goals now tilting away from putting the sciences in the service of the military, and toward the promotion of science, the servicing of industry, and the securing of funds for academic science (Hintz 2013, 47). Even though the NRC initially struggled to redefine its mission, it eventually came to act somewhat like a trade association lobbying for the sciences (Mirowski & Sent 2008, 648). In the 1920s, its main purpose became the coordination of research between the government, industry, and academia (Bugos 1989, 2), especially fostering relationships between private patrons and universities. Industrial research played a major role in these schemes, as is illustrated by one of Hale's measures to achieve his goals of supporting research: the Industrial Advisory Committee. Founded in 1918 and staffed with – amongst others – George Eastman and Pierre S. du Pont, its sole purpose was to lend prestige to a "propaganda program for science" (Kevles 1968, 435; Davis & Kevles 1974, 209). Struggling for organizational reasons, the Committee was reanimated in 1923 with the creation of the Division of Engineering and Industrial Research, led by Jewett of AT&T/Western Electric (Cochrane 1978, 288ff), and subsequently launched a campaign to "extol the virtues of industrial research to industrial executives and the general public" (Hintz 2013, 47). The Industrial Advisory Committee can also be seen as a precursor to another organization that boosted the arsenal of science in the transformations following WWI: the National Research Endowment (NRE, later renamed to National Research Fund).

To summarize, similar to militarization, the politicization can be interpreted in two distinct layers: agency and organization. Not only did professional chemists become active in the political field, lobbying for protection and funding, but also chemistry as a discipline found new patrons. These patrons were not only industrial in nature, as laboratories and fellowships undoubtedly existed before the war, but also military agencies and newly founded organizations such as the NRC, which – as a quasi-governmental entity – brought the needs of science and chemistry into the heart of the political field. Many of these organizations built bridges between academia, industry, and politics, put important proponents of industrial research in efficacious positions, and had one common goal at heart: the popularization of science.

Popularization

As illustrated above, WWI transformed the relationship of chemistry – in all its facets, be they at home in academia or in an industrial laboratory – with industry, the military, and federal agencies. Another area that saw significant changes during these years were the relationships between chemistry and the public (Rhees 1987, 72ff; Rhees 1993a, 45). This phenomenon, the popularization of science, and with it chemistry, had its roots even before the onset of mobilization and war. At beginning of the century, several spokesmen for research emerged that would preach the "gospel of industrial research" (Rhees 1987, 15), such as Robert K. Duncan, the director of the Mellon Institute, a private research institution founded in 1913, and Arthur D. Little of the chemical consultancy service. Both attempted primarily to reach industrial circles and to bolster support for chemical/scientific research in industry by publishing books such as Duncan's "Chemistry for Commerce" (Duncan 1907a) and "The New Knowledge" (Duncan 1907b), as well as Little's Journal and newsletter (Rhees 1987, 23ff, 43ff). Interestingly, these early "educational" efforts came from industrial or industry-related chemists, and not academia (ibid., 51).

Similar efforts picked up intensity, reach, and patronage with the beginning of WWI. For the first time, chemistry made national headlines, "due to the publicity generated by the use of high explosives, the dye famine, and chemical warfare." (Rhees 1993a, 45). The chemist was now in the limelight of a public that often confused them with the arcane alchemist (ibid.). Criticism was leveled at the profession, with some writers wondering how the problems of chemical shortages following the embargo were not anticipated (LaFollette 1990, 8; Rhees 1993a, 45). To address their public image, the ACS, the Chemical Foundation, and other organizations started a "massive crusade to popularize chemistry" (Rhees 1993a, 45). Their experiences with the dye tariff campaign, in particular, made the ACS realize the importance of public perception and the need to cultivate relations not only with congress and industry, but also with the media, viz. journalists and editors (Rhees 1987, 65).

Whereas earlier efforts at creating and distributing a popular journal had failed, 1919 saw the establishment of the ACS News Service under the guidance of former ACS president Charles Herty, one of the key actors in popularization efforts. The News Service was seen as a means to "to keep the public informed of chemical developments and more seriously to restore public confidence in chemistry" (Bensaude-Vincent 1997, 323). The bulletins, written not by chemists themselves, but by journalists, were sent weekly to over 900 daily newspapers in an effort to keep the public informed on events in the world of chemistry (Thackray et al. 1985, 79). With the foundation of the News Service – the first publicity service of any American scientific society (Rhees 1993a, 45) – a new era of science popularization had begun, wherein large institutions became active in the communication of science to the masses (Rhees 1987, 249). Next to these and similar organizations, individual spokesmen were also very active in the popularization efforts, who together "produced cascades of celebratory literature" (Thackray et al. 1985, 100f). Seminal works include Slosson's "Creative Chemistry" (Slosson 1919), and Hendrick's "Everyman's Chemistry" (Hendrick 1917; Rhees 1987, 212). Slosson's book became a central element of the Chemical Foundation's dye tariff campaign, and the organization served as one of the main funding and disseminating agencies of popular literature on chemistry (Rhees 1993a, 46; Thackray et al. 1985, 101). The activities of Garvan, the head of the Foundation, and the other chemical popularizers came to be known as "chemical boosterism" (Thackray et al. 1985, 101). The dye tariff campaign, in particular, indicates how, with the beginning of the 1920s, the gist of popularization shifted from underscoring the importance of chemistry to national defense, and toward the necessity of chemistry for economic progress. Concurrent with the need for chemistry, a need for industrial research was communicated by Maurice Holland, head of the Division of Engineering and Industrial Research within the NRC. Holland's plan included "bringing about an appreciation of the value of research to the public, and popularizing the results of research which will reach the man in the street" (Holland, quoted in Hintz 2013, 47). But Holland did not just aim at the public, but at the industrial scientists themselves, clarifying the necessity to translate their research into understandable, appropriate terms (ibid., 48). Similar to the Chemical Foundation's use of books, the NRC published volumes to support their claims of the wonders of industrial research, such as Holland's "Industrial Explorers" (Holland & Pringle 1928). As can be seen by the variety of organizations, actors, and communication channels, no definite "genre" of science popularization existed following WWI (Bensaude-Vincent 1997, 330), with the chemical boosters utilizing a variety of means to reach and influence industry, politics, and the public.

Once again the deep interconnectedness of the layers of politics, the military, and industry is evident in the popularizers' plight. Popularization being a reaction to the public perception of shortcomings in the chemical industries, or an eminent part of the dye tariff campaign that brought together industrial and academic interests in their efforts to lobby Congress. Even though it was seemingly personified in only a few pivotal figures, the "crusade" had many constituencies and could only be successful by touching the ground with federal agencies and industrial patrons, as well as journalists and the public. Certainly, the campaign to enhance the public image was successful, not only for the increased perception of chemists as professionals but also for the chemical industry itself (Rhees 1993a, 46). This can also be interpreted as a sign of the successful maturation of chemistry as a discipline in the United States, having been transformed from a "marginal branch of arcane knowledge into a key national resource" (Rhees 1987, 347). It remains to be seen how the efforts to alter the reputation of chemistry and those preaching the "gospel of industrial research" were linked – most likely it was a complex relationship regarding questions of where research should take place, as well as the boundary struggles of academic and industrial chemists.

All the efforts to alter the public perception of chemistry need to be seen in relation to the greater backdrop of the popularization of science in general, which changed the position of science in American culture. While not anchored within American life around the times of the Civil War, science came to be seen as the machine fueling progress and prosperity in the following decades, further boosted and complicated by the events of the Great War, with many scientists increasingly finding their way into popular culture. Science popularization is not a linear process, but a complex, multidimensional phenomenon drawing upon a variety of sources: "The various images of science are never definitively outmoded or obsolete. Like the layers of a collective imagination, they can be reinvigorated according to circumstances" (Bensaude-Vincent 1997, 336). Now that we have charted the organizational field of chemistry and its multitude of interrelations and layers, it is these residual, lingering images of chemistry, of science, and of culture that need to be discovered and accounted for discursively when trying to understand how the growing clout of science legitimized and shaped the industrial laboratory.

Notes

1 For a comparison of ACS, AIChE, AES, and AIC membership over the years, see Thackray et al. (1985, 185, figure 6.4-1).

2 As Haynes (1954a, 401na) emphasizes, the early years up to 1910 need to be treated with care, as classifications of "general chemicals", which the number of establishments is based on, changed repeatedly. Furthermore, "general chemicals" exclude sulfuric-nitric acids and other industries such as dyestuffs or fertilizers.
3 As Lazby (1966, 262ff) argues, the relationships built between scientists and the military were only temporary and vanished during demobilization and ensuing isolation, with the main outcome of the war being organizations such as the NRC, scientists' activities in international scientific organizations that sought peaceful cooperation, as well as "exertions to enhance the prestige and value of science at home" (Lazby 1966, 263; cf. also Dupree 1957, 324f).

5 Capturing the Discourse about Industrial Research in US Chemistry, 1870–1930

Methodological Considerations

The relevancy of the concept of discourse to the study of the diffusion of innovations, as well as to the legitimation and institutionalization of new organizational forms, was introduced in Chapter 2. While our theoretical model covered what discourses are – bodies of texts – and what they do – construct and structure social reality – the goal of this part is to get a methodological handle on the concept and to illustrate its operationalization in relation to the organizational field of chemistry. First, some general considerations as to the varieties of and approaches to discourse analysis will have to be made, followed by a discussion of organizations as "speakers", or text-generating entities in the field of US chemistry. Lastly, methodical issues and heuristics for the assembly of a corpus of texts will be discussed.

How to Capture Discourse and Where to Start

The question of how to capture discourse puts a few problems in the way of the intrepid discourse analyst. The notion of discourse always exists in the area of tension between discourse theory and discourse analysis as a method, resulting in a series of theoretically informed clues that aid in operationalizing the concept. One of the central tenets of discourse analysis is that discourses simply cannot be found in their entirety or studied directly, but only through texts that serve as indicators or material manifestations of a larger discourse (Chalaby 1996, 688; Phillips et al. 2004, 636). A text can be regarded as the basic unit making up a discourse, with a relatively loose underlying definition of "text": It includes not only written material but also any symbolic expression with a material component, such as written documents, artwork, recorded words, pictures, symbols, or other artifacts (Phillips et al. 2004, 636). But texts are not meaningful individually, "it is only through their interconnection with other texts, the different discourses on which they draw, and the nature of their production, dissemination, and consumption that they are made meaningful" (Phillips & Hardy 2002, 4; cf. also Chalaby 1996, 687; Wodak 2001, 66; Schmidt 2008, 310). So to perform a discourse analysis – to capture a specific discourse – one has to analyze bodies of interrelated texts.

Assembling a selection of texts into a corpus to be analyzed is one of the first efforts of a discourse analysis, as well as a methodical task without clear-cut instructions. There exists a large variety of ways to conduct discourse analysis, all informed by research traditions of a similarly large variety of disciplines (van Dijk 1985a, 1997; Potter & Wetherell 1987; Keller et al. 2010; Hyland & Paltridge 2011). Variations of discourse analysis such as genre analysis, narrative analysis and critical discourse analysis or approaches informed by the sociology of knowledge, the analysis of hegemony, or more formalized procedures relying on simple coding schemas all offer their own ideas about assembling a corpus and analyzing texts (cf. Phillips & Hardy 2002, 18ff; Phillips et al. 2004, 637; van Dijk 1985b; Wodak & Meyer 2001; Klincewicz 2006). Inevitably, every discourse analysis is unique, depending on the theoretical emphasis and the choice of methodical toolbox – decisions necessarily informed by research question, site, and availability and type of material (compare the different types of texts collected and the different ways they are used, although starting from the same theoretical premises; Munir & Phillips 2005; Maguire & Hardy 2006, 2009, 2013). As Keller (2010, 218) argues, neither the criteria for the selection of texts nor the textual analysis can be completely standardized (Schwab-Trapp 2010, 171f). Accordingly, any serious discourse analysis needs to make all those decisions involved in the assembly and analysis of a corpus of texts as transparent, documented and traceable as possible, presenting well-founded reasoning at all of the critical junctures that come up when faced with the tasks of selecting or omitting texts, developing heuristics for analysis, and generalizing interpretation.

From Field to Speaker

These preliminary observations illustrate the lack of clear-cut, step-by-step guides on conducting discourse analysis. Which brings us back to our initial question – Which texts should be looked at to understand the discourse surrounding the early industrial research laboratories? As outlined earlier, the research site (Phillips & Hardy 2002, 66ff) of this study is the organizational field in which the early R&D laboratories flourished. Utilizing the notion of organizational field aids in steering clear of preconceived notions of academia or industry and their respective organizations, allowing the assembly of a hybrid field of organizations at the intersection of academia, industry, and government. The organizations described in detail in Chapter 4 are considered "speakers", that is, text-generating entities contributing to the discourse on industrial research laboratories (cf. Keller 2010, 210). Of course, not all of the organizations found in the tables of the preceding chapter generate discursive texts of relevance or at all. Other reasons for trimming the list of possible speakers exist as well: some organizations were considered too local or regional such as the NY Section of the Verein Deutscher Chemiker, whereas others were eventually subsumed under the umbrella of larger organizations such as the ACS. Again, others, like the American Association of Cereal Chemists, were upon scrutiny considered of minor interest to the organizational field that saw the genesis of the R&D laboratory. Furthermore, those organizations founded during the war, in particular, brought together authors from academia, industry, and other fields whose positions were doubtlessly put to print in general journals. In this way, historical knowledge of the organizational field aids in generating a list of speakers, particularly in determining which speakers are relevant and in what publications the perspectives of a variety of less "vocal" speakers may be covered at all.

What kind of texts are these organizations producing? In the case of the ACS, publications such as JACS or the Proceedings can be regarded as texts, with the same holding true for the Journal of the American Section of SoCI or the Bulletin of the NRC, to name but a few examples. These texts can be considered "naturally occurring" (Phillips & Hardy 2002, 70ff), as opposed to researcher-instigated discourse such as interviews. Thus, the organizational field of chemistry offers a rich collection of texts from a variety of sources in the time frame of 1870–1930. But the thick description of the historical development of the field served not only to identify speakers but also to provide context, allowing the construction of a timeline of events and chronicling "who *did* what, and when" (Maguire & Hardy 2009, 153, emphasis in original). As outlined, I am interested not only in the texts themselves but also their conditions of production, dissemination, and consumption – a task only made possible through the context that the historical trajectory of the field provides: "... if we are to understand discourses and their effects, we must also understand the context in which they arise ..." (Phillips & Hardy 2002, 4). The positions and interrelations of text-generating organizations can only be understood if looked at in their historical context, and especially with concern for the entrance of new organizations to the field, as is the case with the AIChE, for example – the relation of its texts vis-à-vis those of the ACS can only be properly assessed when understanding the conditions surrounding its foundation. As Fairclough (1992, 71ff) puts it, it is important to situate discursive texts within a three-dimensional conception of discourse (ibid., 73, Figure 3.1), to place texts within surrounding discursive as well as social practices, and to connect "texts to discourses, locating them in a historical and social context, by which we refer to the particular actors, relationships, and practices that characterize the situation under study" (Phillips & Hardy

2002, 4). Coupled with the event timeline, this added layer of discourse captures "who *said* what, and when" (Maguire & Hardy 2009, 153, emphasis in original). Such a discursive-historical approach can also serve as a triangulation, with the variety of empirical data and contextual information being used to check for analyst biases (cf. Wodak 2001, 65). Similar perspectives on how discursive texts and contextual material inform and supplement each other can be found in studies that use an ODA framework (Lawrence & Phillips 2004; Munir & Phillips 2005; Maguire & Hardy 2009).

Two questions have to be answered before moving on to corpus assembly – How does one make sure the timeframe is adequately covered and no speakers of importance are omitted, either in outlining the boundaries of the organizational field, or in the trimming process of speakers? Also, what about the individual scientists writing texts, giving speeches, and in this way influencing discourse? As to adequate coverage, a wide variety of historical works on the early laboratories, the history of chemistry, and American culture in the post-Civil War period in general (Woodlief 1981) were read and included in the contextualization, as detailed in Chapters 3 and 4. Furthermore, numerous reference works were used. In regard to publications pertaining to chemistry, Bolton's *Select Bibliography of Chemistry* and Crane et al.'s *Guide to the Literature of Chemistry* were used in particular to find additional possible speakers by way of produced texts (Bolton [1899] 1967a, [1903] 1967b, [1893] 2005; Lovelace & Thomas 1919, 1920; Crane et al. 1927; Whitrow 1976), while lists of scientific, technical, and other societies (e.g. NAS 1971) generated organizations that could be possible additional text-generators. To find additional industrial corporations, trade associations, and other speakers possibly overshadowed by the pioneers, publications such as West (1930) and the Annual Records of Science and Industry (Baird 1872–1879) were scoured for details. Information on newspapers, trade, and popular science journals (Mott 1957a, 1957b; Elfenbein 1960; LaFollette 1990) was used to account for more general speakers. Eventually, what can be regarded as "saturation" occurred, leading to what can be considered rather robust coverage of the organizational field from 1870 to 1930. The relation of text to discourse aids in theoretically informing the conclusion of such a robustness check: No text reproduces a discourse in its entirety, but not all texts are needed to properly analyze a discourse. So, while it surely would have been evident during the analysis of the texts themselves if important events or speakers had been omitted, "omission by saturation" of possible minor speakers (e.g. local and regional scientific societies) does in no way impinge on the capacity to analyze the discourse.

Of course, most texts are produced by individual authors and not organizations. But, their ability to produce and disseminate texts is always mediated by organizations. In this way, authorship by Willis R. Whitney becomes meaningful through his position as the eponymous research director of the GE Laboratories, while A. D. Little's paeans of praise for industrial research need to be understood with regard to his position as a famous scientific consultant for AD Little, and eventual president of the AIChE and SoCI, etc. Similar premises apply for the presidents of scientific and technical societies, which are positions warranting "voice", that is, the ability to be perceived as legitimate contributors to the discourse, the resources to be heard, and the centrality to reach many actors in the field (Phillips et al. 2004, 643). Any meaningful analysis needs to take account of the subject positions created and transformed by discourse (cf. van Dijk 1997, 19ff), and these legitimized roles can be attached to individuals (or at least their public personae), as will be discussed in Chapter 6. By situating the research question within a framework of the sociology of organizations, assembling a field of speakers means looking at organizations first and always framing text-production and authorship within the three-dimensional context of the field.

From Speaker to Corpus

The last step after assembling a list of text-generating organizations as speakers of discourse is to collect these texts that are understood as an interrelated body or "corpus" (cf. Phillips & Hardy 2002, 5). Of course, in a discourse analysis of this size – spanning sixty years and a multitude of speakers – choices have to be made, both in regard to the speakers and the selection of texts. Since my research interest was focused primarily on scientific discourse, several types of texts such as company statements and reports were excluded from the analysis to assure a certain comparability of discursive texts – ideas, statements, and comments from scientists and managers of the firms in question would still come up in scientific and trade journals. Such materials were instead used for the further framing of events and trajectories within the field, as indicators for the establishment of laboratories. Contributions of this adjusted field of speakers can be regarded as a set of overlapping genres, that featured certain conventions for creation, as well as recognizable elements among the texts (Tardy 2011; cf. also Fairclough 1992, 232f; Phillips et al. 2004) – resulting in "a kind of tacit contract between writers and readers, which influence the behaviour of text producers and the expectations of receivers" (Hyland 2011, 174f). Since the goal of this analysis is not genre but the constructive functions of discourse, corpus texts will not be subject to a full-fledged genre analysis. On the one hand, what the notion of genre adds, though, is a certain sensitivity to the conventions of (academic or popular) discourse and how they were followed or transgressed within individual texts – practices that can influence the visibility, reception, and impact of texts (Phillips et al. 2004, 643f). On the other hand, it serves as a further frame of reference for locating the discursive corpus within its historical context.[1]

As to the choice of choosing "only" journal publications, and not books or other types, in addition to the conception of journal publications as a genre, field-level and pragmatic concerns played a part in this decision. Since our point of access to the field is embraced within the speaker, that is an organizational entity, it was sensible to utilize the normal avenue for communication of these organizations, which comprised journals in which their reports of meetings, announcements, and current discussions were published. The importance of certain books for the science popularization movement was related above – such books and other discourse-changing works, while not part of the discursive corpus, still found entrance in the analysis through intertextual relations (e.g. book reviews; see later) established within texts, making it unlikely to miss important non-journal texts having an impact on the discourse's trajectory. Moreover, the higher cadence of journal publishing and the nature of events in question made it more likely to be covered in such outlets, rather than books which would take much longer to prepare and be received.

Since analyzing thousands of text would be a fruitless endeavor, both from a theoretical and a practical standpoint, a search heuristic was developed to generate a reasonable corpus, utilizing a list of keywords that was informed by the stories of the pioneer laboratories and the happenings in the field (Table 5.1). While some search terms were quite self-evident, such as "research laboratory", others, for example, "chemical engineering" or names of specific organizations were added after the history of the field was mapped and their importance to the history of the R&D laboratory became apparent. Search terms were applied to the titles, and – if available, which rarely proved to be the case – the abstracts. In some cases, journal issues could be searched automatically if title information was extracted in a machine-readable format, like for most ACS journals. In others, the table of contents or sometimes entire issues had to be manually scanned for relevant texts. After titles and abstracts, author names were used as the last iteration of corpus assembly. Authors' relevance was informed both by the field analysis, as well as their previous textual contributions – if authors turned up that had repeatedly written texts deemed important and fitting to the discourse, attention was devoted to find other publications by those authors. Certainly, this course of action is prone to random or systematic errors. Random errors can, of course, never be fully ruled out, although attempts at curtailment were made by carrying out the corpus assembly in iterative steps and re-checking select issues. The more problematic dangers that can arise from systematic errors were prevented through the historic-discursive approach of contextualizing the results, as well as thorough checks for the robustness of search terms.[2]

Table 5.1 Search Terms Used for Corpus Assembly

Search term(s)	*Description*
GE Laboratory, Du Pont Laboratory, AT&T Laboratory, Kodak Laboratory, and iterations	Various names and descriptors for the pioneer laboratories, also their respective subsections (Departments).
Science	Including variations such as "science-based", "scientific", "pure/applied science", "scientist", "man of science", etc.
Research	As a verb and noun, in variations, e.g. "researcher", "research institute", "research laboratory", "fundamental research", etc.
Laboratory	In variations, e.g. "chemical laboratory", "metallurgical laboratory", "laboratory methods", etc.
Invention	In variations, e.g. "inventor", "inventive", etc.
Chemistry	Including variations such as "chemical", "chemistry-based", "chemist", "chemical analyst", "industrial chemistry", "chemical industry", etc.
Chemical Engineering	Including variations such as "chemical engineer", "engineering chemistry", etc.
Industry	Especially in relation to terms such as "chemical", "science-based", etc.
Proceedings, Annual Meetings, Editorials, …	Especially in early years, special attention was paid to Proceedings/Meeting Reports of organizations as they may offer glimpses of organizational self-description and current discussions surrounding that organization, as well as presidential addresses covering current topics of note.
Organization Descriptors	Names of organizations, e.g. ACS, AES, but also others making up the research landscape such as the Carnegie Institution. List of organization descriptors grew iteratively.

Dataset Overview

Figure 5.1 shows both the number of articles per year and the number of journals analyzed. Subject to some minor gains and losses in the early years, the number of journals steadily rises as the discipline of chemistry specializes and differentiates with new organizations and subdivisions of the ACS (cf. Table 4.2) – tripling from an initial eight in 1870 to 24 in 1930. As can clearly be seen, the amount of articles published that were identified using the search heuristic detailed previously grew rapidly with the beginning of the twentieth century, reaching a peak around WWI and the postwar reconfigurations of science and industry, and totaling 2992 articles by 1930.

In order to give an overview of the journals included in the dataset (Table 5.2), they were grouped by organizational affiliation. Any analysis of this kind, being based on an organizational field that brings together many different kinds of speakers with varying visibility, resources, and legitimacy (Phillips et al. 2004, 643), needs to account for these differences. Yet one runs into the grave danger of letting the analysis of texts be guided by preconceived notions of both position and content. In practice, this would take the form of assuming any author publishing in JACS would be an academic scientist supporting ideas of "pure science" as opposed to applications of science in industry. Evidently, such an ordering of discursive texts would greatly influence the results, most likely reproducing presupposed ideas about the relationship between

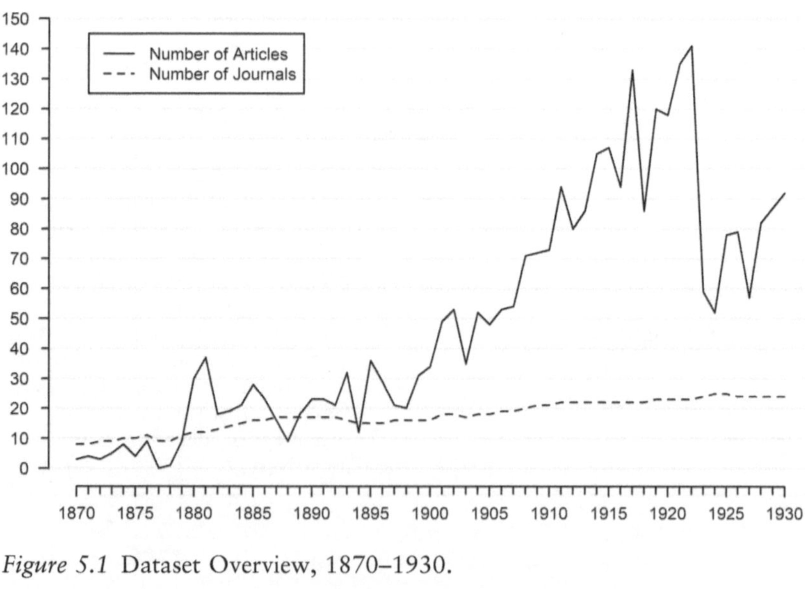

Figure 5.1 Dataset Overview, 1870–1930.

Table 5.2 Dataset Overview

Journal	*First year*	*Last year*	*Access*	*Notes*
American Journal of Science (& Arts)	1870	1930		
Journal of the Franklin Institute	1870	1930		
Journal of Applied Chemistry	1870	1875		
Proceedings of the AAAS	1870	1930	1911	
Proceedings of the American Academy of Arts & Sciences	1870	1930		
Scientific American	1870	1930	1922	
American Chemist	1870	1876		
Nature	1870	1930		
Popular Science Monthly	1872	1930		
The Laboratory	1874	1876		
ACS Proceedings	1876	1930	1920	
American Chemical Journal	1879	1914		1914 merger with JACS
JACS	1879	1930		
Science	1880	1930		
JSoCI	1882	1930		
Bulletins of the Division of Chemistry, USDA[a]	1883	1902		From 1885 onward in Bulletin of Association of Official Agricultural Chemists
Bulletin of the Chemical Society of Washington	1884	1895		From 1893 onward local section of ACS
American Analyst	1885	1892	1891	
Journal of Analytical Chemistry	1887	1893		1893 merger with JACS
Journal of Physical Chemistry	1897	1930		
Electrochemical Industry[b]	1902	1930	1922	

(*Continued*)

Table 5.2 (Continued)

Journal	*First year*	*Last year*	*Access*	*Notes*
Transactions of the AES	1902	1930	1922	
Chemical Engineer[c]	1904	1925	1922	
Journal of the American Leather Chemists Association	1906	1930	1922	
AIChE Transactions	1908	1930	1922	
Journal of Engineering and Industrial Chemistry	1909	1930		
Chemist-Analyst	1911	1930	1922	
Journal of the Association of Official Agricultural Chemists	1915	1930	1922	
Bulletin of the NRC	1919	1930	1922	
Chemical & Engineering News	1923	1930		
Journal of Chemical Education	1924	1930		

Notes

a The Bulletin was published independently at first, from 1885 in the Bulletin of Official Agricultural Chemist. It changed its name to "Bulletin of the Bureau of Chemistry of the USDA" in 1901.

b Formed as "Electrochemical Industry", this journal underwent name changes: "Electrochemical and Metallurgical Industry" (1905–1909); "Metallurgical and Chemical Engineering" (1910–1917); "Chemical and Metallurgical Engineering" (1918–1925). Furthermore, "Iron and Steel Magazine" merged with this journal in July 1906.

c Merged with "Chemical Age" in 1920 and continued publication under the new name until 1925.

"First Year" denotes the first year the journal became part of the analysis (not first year of publication); "Last Year" denotes the last year the journal was included in the analysis; "Access" refers to issues of accessibility, some journals were published over the whole period yet no access could be gained to later editions.

Removed from analysis:

- Oil, Paint and Drug Reporter (1903-unknown): more relevant to pharmacy than chemistry
- Chemical Week (1914-unknown): no access options could be found
- The Chemist (1923–1930), Journal of the AIC: no physical/digital copies prior to 1930 could be located.

academia and industry in the process. Similar problems would arise if categorization were based on authors' affiliations. In many cases, an author's affiliation was simply a city, only later being replaced by organizational affiliations with universities, corporations, and government agencies. But even the more precise details of the later years run the risk of omitting important aspects in the case of multiple places of affiliation, e.g. a chemist at university also employed as a consultant. As Keller (2010, 221) emphasizes, analysis according to fixed attributions must be avoided. Thus, the proposed ordering of journals – as speakers of discourse – makes no assumptions as to the content of texts, instead attempting to group speakers according to similar places of publication in the field. These places of publication allow the highlighting of conditions of production as well as possible audiences (cf. Fyfe 2016, 395ff). Moreover, it is important to emphasize that the categorization presented here was merely an analytical tool, used to further reduce and order the size and volume of discourse by introducing rough distinctions that were derived from informed theoretical assumptions about the nature of the field and probable lines of argument within, and made possible through the historic-discursive approach.

Popular Journals

Two journals that can be considered popular science journals were included in the analysis: Scientific American (1870–1930) and Popular Science Monthly (1872–1930). As LaFollette notes, both journals can be considered "science magazines intended for audiences attentive to science" (LaFollette 1990, 24). Scientific American was a widely read magazine with a circulation of around 40,000–50,000 by the late nineteenth century, well-known for its announcements of new inventions and a weekly official list of patents. Publishing "a variety of information of mechanical and scientific nature" (Mott 1970a, 318), the focus on invention and patents gradually ceased by the twentieth century's second decade, making way for articles of a broader scope and communicating news of the sciences to the people: "It was no longer the inventor's paper, but a periodical of popular science" (ibid., 323). Popular Science Monthly was founded with the intention of bringing news of scientific progress in Europe to American readers who had no scientific training (Mott 1970b, 496). Over the years, it shifted toward American science, reprinted addresses held at learned societies, and featured articles by college and university professors (ibid., 497). A change in ownership in 1915 brought further adjustments, with the journal now presenting scientific information in a "definitely popular fashion", geared toward hobbyists (ibid., 499). The inclusion of these two journals can of course in no way account for the entire field of popular science publishing, which is not the goal of this analysis. Instead, by including these two journals that were arguably two of the widest circulated popular science magazines, the "popular" reaction to the advent of the R&D laboratory – if any such reaction happened – is added to the analysis, giving further context to ideas on science, chemistry, and laboratories that were constructed and spread by academic and industrial scientists.

General Science Journals

In order to assess the level and magnitude of discussion, especially in the early years when specialized chemical societies and journals did not yet exist, several publication organs of important organizations that played a part in the professionalization of chemistry were included. The Proceedings of the AAAS (1870–1930) and the Proceedings of the American Academy of Arts & Sciences (1870–1930) were included to map discussion in scientific societies devoted to the sciences in general. Especially with the founding of AAAS Section C in 1882, texts regarding the nature of the chemist, chemistry, and so on were expected, yet neither journal yielded many hits when scanned for relevant terms. This is cushioned by the inclusion of Science (1880–1930) in the analysis, the journal acting as the house organ of the AAAS from 1900 onward. It arguably became the national periodical of the scientific community after its editorial reorientation in the late 1890s, covering a wide variety of sciences and serving as a forum for disputes and controversies (cf. Vandome 2013). This journal made up for a large part of the general science publications analyzed, covering much of the disciplinary and extra-disciplinary debate on the early laboratories, the professionalization of the chemist, and the applications of science. The British journal Nature (1870–1930) was added, since cursory analysis showed frequent references to articles published in Nature, a finding further bolstered by the fact that in the early decades, American discourse may have lacked places of publication due to a lack of professionalization and specialization. In turn, many authors went to foreign journals instead, especially Nature, which served as a guide for emerging American journals of science (Vandome 2013, 174). Nature also played a seminal role in the process of defining what being a scientist meant during the ascent of the scientific community to a position of cultural authority (Baldwin 2015), a process that doubtlessly played an important part in the emergence of industrial laboratories. To round out the picture, the American Journal of Science (1870–1930) and the Journal of the Franklin Institute (1870–1930) were included, the former due to its position as one of the earliest and most important outlets for scientific writing in the United States, and the latter as the organ of a well-respected scientific institution documenting scientific and technical progress (Mott 1957a, 556ff), with both yielding only a few texts of interest.

Chemistry Journals

Journals published by chemical societies or individuals explicitly positioning their publications within academic chemistry make up the largest group within the analyzed corpus. In the years prior to the founding of the ACS, as well as during its difficult first years, many short-lived publications served as homes for chemical thought: the Journal of Applied Chemistry (1870–1875), the American Chemist (1870–1875), The Laboratory (1874–1876), and the American Analyst (1885–1892). When the ACS entered the scene, it did so by first publishing its Proceedings (1876–1930), followed by JACS (1879–1930). JACS absorbed numerous small journals over the years, such as the American Chemical Journal (1879–1914), the Bulletin of the Chemical Society of Washington (1884–1894), and the Journal of Analytical Chemistry (1887–1893). The publication landscape was heavily influenced by ACS publications, such as the Journal of Physical Chemistry (1897–1930), until the founding of further specialized societies and their organs: the Transactions of the AES (1902–1930) and the AIChE Transactions (1908–1930). Industrial chemists received a voice with the establishment of their respective subdivision in 1908 (Journal of Industrial & Engineering Chemistry [JIEC], 1909–1930). After the war, further specialization, as well as increased scientific communication, is evident from the foundation of journals such as Chemical & Engineering News (1923–1930), sometimes called the "newsweek of the chemical world" (ACS 1951, 20), and the Journal of Chemical Education (1924–1930). As outlined earlier, local and regional sections of chemical societies were not included – or only in the early stages of the period – since it can be assumed that the national discussion that was condensed in larger publications will adequately serve as an indicator of discursive topics and shifts. Regarding the journals chosen for analysis here, organs communicating meetings and the current happenings in the field – Proceedings of the ACS, Transactions of AES/AIChE – make up the bulk of the texts selected.

Industrial Journals

Among journals published by speakers that can be considered "industrial", such as corporations or trade associations, the Journal of SoCI (1882–1930) plays an important role, especially with the inception of the New York Section (1894) of the British society, and presidential addresses that frequently discussed the achievements and needs of the American chemical industry. After 1900, several journals devoted to the applications of various branches of chemistry in industry were formed: Electrochemical Industry (1902–1930), and the Chemical Engineer (1904–1925). Whereas journals such as the Chemist-Analyst (1911–1930) specifically targeted chemists working as analysts employed industry. Furthermore, the associations of chemists in specific fields such as leather chemistry or agricultural chemistry published their own periodicals: Journal of the American Leather Chemists Association (1906–1930) and Journal of the Association of Official Agricultural Chemists (1915–1930).

Governmental Journals

As the last group, two journals that originated in government agencies were included to enlarge the field of possible speakers and account for the important roles that both agencies played in the development of chemistry as well as the granting of visibility and status to the chemist. The first of these are the Bulletins of the Division of Chemistry, published by the USDA (1883–1902). As outlined earlier, the USDA proved to be an early and important employer of chemists, especially after the Morrill Act and the founding of the land-grant universities. Also included was the Bulletin of the NRC (1919–1930) in an attempt to map the changed attitudes toward chemistry and the chemist in the wake of the "chemist's war". As it turns out, while the USDA Bulletin added no relevant texts to the corpus, the Bulletin of the NRC proved to be a place of discussion regarding the new functions of research and its place within a newly altered landscape of science in post-war United States – a discussion that is fruitful and highly relevant to the discursive concepts and shifts that are being sought here.

An Event-Based Analysis

Similar to the steps that were necessary to make sense of the organizational field and to order and reduce the number of speakers to a manageable amount, further condensations for the identified texts are necessary. No clear-cut recipes for theoretical sampling of this kind exist (Phillips & Hardy 2002, 74), so the decision was made to analyze the data based on events that were identified as relevant during the construction of the organizational field (cf. Keller 2010, 214f). Events that challenge existing structures and impact the field in a variety of ways can be understood as socially constructed since they are given meaning discursively (Munir & Phillips 2005, 1669). In this way, events are problematizations, which leads to discursive texts trying to make sense of changes or strategically promoting new concepts or subject positions: "Actors inscribe problematizations of institutionalized practices in texts in order to allow these ideas to travel in space and time" (Maguire & Hardy 2009, 151). Events are chosen with regard to research interests: those that may have caused ruptures and shifts in the discourse on chemistry and the place of science. Approaching a longitudinal discourse corpus not as a whole, but at several points in time, also offers the advantage of comparing shifts in the construction of meaning between multiple events:

> Historically oriented analyses can proceed by conducting several synchronic cuts through a discourse strand – based on discursive events, for example – and subsequently comparing them with each other. Such analyses provide information on changes to, and continuities of, discourse processes through time.
>
> (Jäger 2001, 52)

Due to the continuous nature of the events, as well as the corpus being made up of journal publications, it was decided to cover not only the year of each event but also one year before and after it. This further reduction of the corpus yielded seven three-year-tranches centered around events chosen for their theoretical and historical relevance – twenty-one years were analyzed in total. In this way, it was possible to include texts immediately leading up to an event, as well as reactions – since writing and publishing takes time, and issues of journals appeared at fixed intervals, for example, monthly or bi-annually, instead of instantly following an event.

Events as Focal Points in Discourse

As Table 5.3 shows, seven events were chosen for the analysis of the discourse. The first, in 1895, concerns the professionalization of chemistry. Of course, as professionalization is a process and not a singular point in time, it is hard to pinpoint a specific year for this. Still, it can reasonably be assumed that the professionalization of chemistry as a scientific discipline, and a field of employment, showed stable patterns in 1895. The ACS had established itself as the major organization for chemists in the country after its tumultuous years, and the early structures for publishing were in place by then. Similarly, several organizations devoted to chemists in industry were in place. Chemists found employment at chemical establishments in a variety of industries, exhibiting steady growth in the 1890s. Hence, it can be assumed that the role of chemists as scientists, and as industrial chemists, was soundly in place. Moreover, in order to understand if and how the discourse changed with the establishment of the early R&D laboratories, it is necessary to capture the discourse before the first laboratory. This event thus serves to outline discursive concepts, objects, and subject positions before the pioneering industrial laboratories.

The second event is the establishment of General Electric's famous laboratory in 1900. As the history of the early laboratories has shown, GE's research program served as a reference point for many research directors in the following years, with the laboratory having acquired an

Table 5.3 Discursive Events

Event	*Year*	*Description*	*# of articles for year*	*# of relevant articles*	*# of articles for year ± 1*	*# of relevant articles for year ± 1*
Professionalization of Chemistry	1895	Stable patterns in the field of chemistry	35	25	76	31
GE Laboratory	1900	Establishment of GE's laboratory	35	22	115	34
DuPont Laboratories	1902	Du Pont laboratory as the first lab in chemical industry	55	28	139	47
ACS Division Industrial & Engineering Chemistry	1908	Concerns of industrial chemists become so pressing that ACS establishes a new division, journal	70	30	194	60
Kodak Laboratory	1912	Establishment of Kodak's laboratory, often seen as the last of the pioneers	80	38	256	84
World War I	1917	Mobilization, dye shortage, transformation of US science landscape	133	47	313	107
Directors of Industrial Research	1922	Establishment of regular meeting of R&D laboratory directors	142	63	335	107
Σ			543	253	1419	470

archetypal position in the discourse. The GE laboratory became an important part of the organizational (and with it the discursive) field especially because of Langmuir's chemical work, even though GE was primarily active in the field of electrical engineering.

1902 brings the third event: the founding of Du Pont's laboratories, both the Eastern Laboratory and the Experimental Station. Following a re-organization, these two laboratories were built within different parts of the company, overcoming different oppositions and taking on different paths. Since these laboratories were the first of the pioneering laboratories to have been established within the chemical and allied product industries, this serves as an important event for observing what changes, if any, the discourse may have undergone since the founding of GE's research department.

The next event, the establishment of the ACS's Division of Industrial & Engineering Chemistry in 1908, showcases interesting transformations in the discourse, especially with regard to the role of the chemist. The concerns of industrial chemists, who felt they were not heard within the existing structures of the ACS, prompted the ACS to establish a new Division, illustrating that industrial chemists as a constituency were significant enough by this time to warrant a reaction by ACS leadership out of the fear of an organizational split. The Division also published its own journal in 1909. Moreover, several other divisions were founded in 1908 and 1909, showing an increased specialization and differentiation within chemistry – trends that had led to the formation of new organizations (AIChE, AES) only a few years prior. This event may serve as a pivotal point for the re-definition of the identity of the industrial chemist, as well as their relation to academic chemistry.

The fifth event concerns the last of the pioneering laboratories to be built: Eastman Kodak's laboratory in 1912. It can reasonably be assumed that the organizational form of the in-house research laboratory, both as a concept and an object, had become perhaps not established but at least known within the organizational field. Kodak's history also shows the pressures to do research that were apparent by the second decade of the twentieth century. The texts that were produced before, and as a reaction to, Kodak's laboratory will further serve to deepen the understanding of the discourse surrounding the new organizational form.

By 1917, the United States had entered WWI – the sixth event. The years before saw the dyestuffs shortage, an industrial preparedness campaign of major proportions, as well as the mobilization of chemists from both academia and industry to serve in the armed forces. These transformations will have impacted the discourse significantly, but WWI also caused a transformation of the organizational and institutional landscape of support for science, with new networks of cooperation and communication between the military, academia, industry and a host of other groups leading to a certain interpenetration of spheres that were previously distinct and undisturbed. It can be expected that both the role of the chemist and the powers of the laboratory had been explicitly and repeatedly problematized in discursive texts produced during this event.

The seventh and last event surrounds the founding of DIR in 1922. This meeting place for directors of research laboratories was launched in 1923. Similar to the 1912 event, it showcases a possible normalization of the research laboratory, given that new organizations for its administration and improvement – granting visibility as well as structures for communication – were founded in its wake. In addition, demobilization was over by then, with the industry having scaled back to prewar levels and chemists increasingly being let go. The new organizational landscape of science funding was firmly in place, and the popularization campaigns signing the praises of chemistry were in full swing. Chemists had also become active politically, pushing for legislation in dye tariff matters. Thus, this event serves to check what discursive changes the war left in its wake and who the chemist was before and after.

Next, these seven three-year-tranches were analyzed to get a "cleaned" set of texts for each year. The main reason for exclusion from the discursive corpus was duplication. As it turned out, a wide variety of texts – such as addresses, announcements, reports of meetings, and comments on policy and the army draft for the latter two events – were published by numerous journals. In this way, the addresses of the presidents of the ACS, SoCI, and AIChE were communicated within their respective house organs, but reprints also frequently found their way into Science or the Chemical Engineer, and announcements of societal meetings were spread far and wide. While removing instances of duplication reduces the discursive corpus by a great degree, they do attest to the visibility of texts and once again punctuate the hybrid nature of the field, thereby strengthening the mapping of the field as an organizational field instead of as distinct spheres of academia, industry, and government. Another reason for exclusion was what may be called "not relevant", which is related to the lack of abstracts and the way the search heuristic works. Often, article titles such as "The American Chemical Society" would lead to inclusion within the corpus, yet upon further inspection, the articles revealed nothing but a few sentences announcing the next general meeting, or a list of papers read at a past meeting. Furthermore, some texts were excluded because they were not concerned with the guiding research questions of this analysis. Such texts may have been either misleadingly titled or chock-full of technical information with little value added to the discourse, such as with market reports and the like. Using these seven events and their respective years prior and after to construct seven three-year-tranches, a total of 1419 articles were left in the corpus, of which 470 were deemed relevant and usable for analysis.

These 470 articles were written by 251 unique authors.[3] Charting the distribution of publications per author, it is immediately evident that there are few authors overrepresented in the discourse that this corpus attempts to capture.

Two authors manage to enter the discourse with double-digit publications: A. D. Little, of consulting and science popularization fame, and E. F. Roeber, who authored thirteen discursive texts by virtue of being the editor of Electrochemical Industry and entering the corpus through numerous editorials. Both Little and Roeber author discursive texts for the first time during the fourth event, in 1908. Even though Little and Roeber manage to repeatedly contribute to the discourse by way of their positions within the field, they seem not to have fundamentally guided, influenced, or precluded discursive constructions. Apart from these two outliers, the absence of frequently published authors indicates both the lack of central figures around whom the discourse revolves, and those who – by nature of their visibility and volume in discursive texts – managed to influence the construction of concepts, objects, and subject positions (cf. Phillips et al. 2004, 643ff). In-text references to people – be they famous scientists, directors, or others – mainly point toward extra-discursive figures, and not the other authors within the corpus (see later).

Breaking the corpus down by organizational speakers reveals a distinct picture (Figure 5.2). Texts published in popular journals remain a small but steady presence, while articles from speakers in the general science group make up the largest part of discursive texts in the early years, declining toward the later years, and peaking again during WWI.

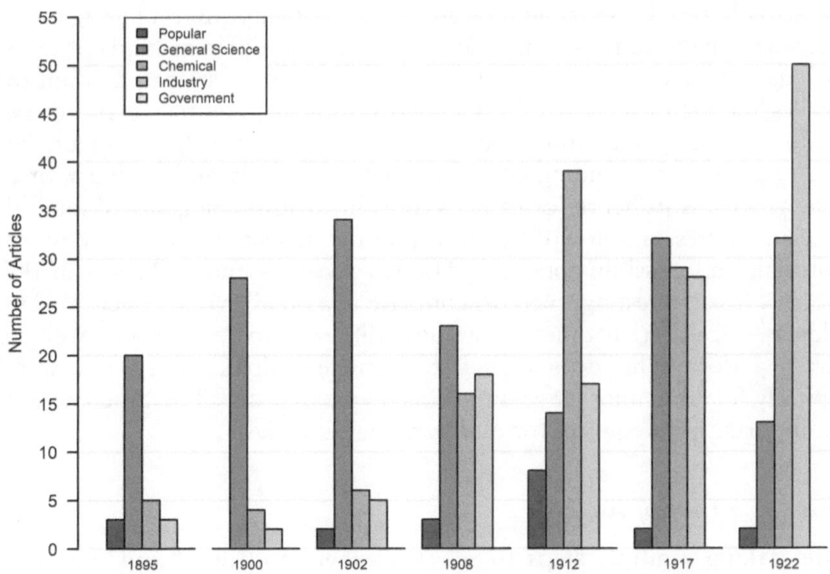

Figure 5.2 Number of Articles per Speaker Group and Event.

Publications by industrial speakers show a strong growth, making up the largest amount of texts by the last event. This trend is mostly due to an increasing number of issues – and with it, articles – from the journals Electrochemical Industry and Chemical Engineer, discussing industrial matters, the place of the chemist, and so on. After the creation of a special division for industrial chemistry and the subsequent publication of a journal, the share of texts written by chemical speakers rises sharply. For the last three events, almost all industrial concerns and discussions are transferred to JIEC, with JACS, the Proceedings of the ACS and other specialized journals hardly contributing to the ongoing discussion anymore. Government speakers initially make up the smallest group, with only two publications included, and do not partake in the discussion at all – their contributions are either not relevant to the analysis or outside of the seven events.

As a last quantitative overview of the corpus, journal articles were roughly categorized into five groups of text types. Of these, articles – ranging from reports on scientific discoveries and industrial developments to treatises on the powers of chemistry and the values of science – make up the largest amount, with addresses at meetings, symposia, and other occasions the second strongest group. The share of editorials – as clearly labeled opinion pieces, sometimes commenting on current events and other times welcoming a new year and reviewing the last – were a growing presence for the last four events. Meeting reports that summarize important happenings and decisions at the encounters of scientific and technical societies, and the papers read at these occasions, similarly increase with the later years, in part due to more organizations holding more meetings. The residual group, "Other", comprises different kinds of texts such as Reviews or Letters to the Editor, representing a small amount of articles for each event. Again, a categorization such as this shall not be used to infer anything about the content of a text; instead, it is an aid for reaching an understanding of the possible audiences and visibility of a text, as well as its genre (or rather sub-genres within the genre of journal article). Addresses follow different rules of composition than a review or editorial, and possibly construct different speakers and audiences in the process. It is interesting, then, to compare texts both within a genre – Do addresses at different times presuppose different audiences? – as well as across genres – What does it mean if an article exhibits patterns that were typically found in another genre? Being aware of such subtle differences is an important prerequisite for any meaningful analysis.

Texts as a Unit of Analysis

After taking multiple steps to select speakers and build and reduce a corpus, 470 texts were left for analysis. The analysis of individual texts in a discourse analysis always exists within the tension of using fixed, rigid categories, veering into the realm of content analysis; and the freedom and creativity that results from the lack of standardized approaches (Phillips & Hardy 2002, 74). The multiplicity of ways that discourse is theorized leads to a multitude of ways that discourse is analyzed, and, as Potter and Wetherell note, analyzing discourse is "like riding a bicycle compared to conducting experiments or analyzing survey data which resemble baking cakes from a recipe" (Potter & Wetherell 1987, 168) – there are no strict step-by-step guidelines, and much of it is learning to get a "feel" for discourse. Some variants focus on the use of linguistic and grammatical constructs in discursive texts, others on larger narratives that can be synthesized from a plurality of texts, and still others on intertextual relations between texts. In the end, it is important to find a style of analysis that fits the type of data, thereby generating meaningful and justifiable categories that relate to the research questions. Due to the scope and intensity of such an analysis, it is equally important to check for theoretical saturation and to know when to stop (ibid., 79; cf. also Jäger 2001, 51).[4]

The research questions guiding the analysis where derived with the aid of the theoretical basis in discursive institutionalism and informed by the historical approach. Questions asked focused on concepts, objects, and subject positions: What is science, what is research, what is chemistry, what is a laboratory, who works in a laboratory, who is the chemist, and the chemical engineer, who is the inventor?, and so on. The set of questions was kept as flexible as possible and open to additions. To relate one example, at a later point in the analysis the relevance of the placement of the laboratories became apparent, thus "Where is the laboratory?" was added to the research questions.

The schema used for analysis was developed through multiple trial readings of discursive texts and guided by several "discursive toolboxes" and exemplary studies (Fairclough 1992, 225ff; Jäger 2001; Phillips & Hardy 2002; Maguire & Hardy 2009, 2013; Keller 2010; Machin & Mayr 2012). As opposed to specifying a wide variety of categories in order to catch all of the layers of meaning, the schema was kept simple on the one hand, and as open as possible on the other. Thus, instead of for example pre-selecting a couple of possible argumentative strategies constructing subject positions and noting their occurrences in texts, the category concerning subject positions was simply used to list roles, and actual people. In the second step, exactly how these roles were constructed and people were referred to were scrutinized, making it possible to let textual meaning and structures guide the process of analysis – meaning and structures which may have been obscured by categories too rigid and preset.

For each individual text, the analysis began by writing a brief summary that could be used for quickly reassessing the contents of an entire text or a single paragraph, with later findings sometimes causing a need for reanalysis by having added attributes to a concept. Next, intertextuality was highlighted (Fairclough 1992, 101ff). References to other texts play an important role in ODA since these textual relations may add legitimacy by leaning on established texts and evoking "understandings and meanings that are more broadly grounded" (Phillips et al. 2004, 644). Intertextuality can take a variety of forms, from explicit citations one might expect in the genre of scientific texts, to passing mentions of books, laws, or current discussions, to the use of recognizable phrasing or terminology associated with specific texts or people, and all the way to implicit allusions (cf. Tardy 2011, 59). Moreover, references can be made to either extra-corporeal texts, that is, texts not included in the discourse analysis, or, of course, texts within the discursive corpus. Frequent references to intra-corporeal texts can give hints as to their visibility and importance for the discourse. As it turned out, intertextuality played only a minor role in the analysis, as many of the texts analyzed were not scientific papers per se, but rather speeches or comments, and the standards for scientific writing were not as established and elaborate as today. Only a few texts made explicit or implicit references, thus few texts within the corpus were elevated to any special importance. Still, the few references made served to add "weight" to the ideas referenced, since intertextual references showed that they were heard and in this way perpetuated discursively, adding to their ability to construct and influence reality.

Third, topics were denominated, in relation to the research questions: Does this text concern itself with chemistry, or the chemist, and does it talk about science or the laboratory? Of course, most texts had more than one topic, which proved to be an emergent category, and topics that later turned out to play a key part in the definition of the laboratory were not part of the original "list", requiring additional classification of earlier texts. Interestingly, for many events there were predominant topics (later understood as other, adjacent discourses) that could be identified – topics such as specialization or the Conservation Movement that were discussed intensely during one event would get a lot less attention before and afterward. Next came the quite openly termed category of "People", which worked in a similar vein as intertextuality, but instead of charting references to texts it searched for people. First, people could mean the authors themselves – Did they position themselves in any way in relation to their subject matter and did they list credentials or experience? Second, people could refer to individuals, ranging from people partaking in current discussions, to research directors of famous laboratories, all the way to famous scientists from the past. Third, people could refer to roles, such as the chemist, the chemical engineer, or the manager and banker. How were they constructed, referred to, and positioned? Like topics, this category required later detailing.

Fifth, events and locations were analyzed. What events, such as societal meetings, scientific discoveries, and laboratory formations, where mentioned in texts, and how where they described? What places were mentioned, such as Washington, DC, Niagara Falls, MIT, and the Kodak Laboratory? In what way were they invoked: Was MIT a place of science, a university, or something else? The next category, "Notable Quotes", was mainly used to ease the workflow of writing up the results, in that telling quotes that underscored or questioned concepts, objects, or subject positions were selected once the analysis of a text was finished. Lastly, "Notes" was used as a residual category where ideas, surprises, and interesting things could be jotted down and early attempts at interpretation or cross-references to other texts could be made. These notes served as starting points in the second round of analysis, giving tips regarding what not to miss, what categories to solidify, and what additional questions to ask. The omission of "concept" and "object", which were central to the interpretation in this discourse analysis, as categories in the schema was made deliberately. Both are highly fragile categories that emerge by (re-)analyzing texts repeatedly – in particular, the materialization of concepts in objects is a complex process that cannot be traced by simply filling out a category in a schema, "checking boxes" for any one single text so to speak. Instead, concepts and objects were identified in the steps following the initial analysis as the effects of readings of multiple texts and the relations their contents constructed.

It is important to stress that working on a text and attempting to synthesize its contents into categories does in no way involve any "reading between the lines" to find out what the author "really meant". Instead, it is about mapping repeatedly occurring statements and probing them for patterns and changes over time. As Foucault (1981) postulated (and then failed to adhere to in his own discourse analysis work, as Schwab-Trapp (2010, 180) notes), discourse analysis is a highly positivist method. This needs to be kept in mind by both the analyst, as well as the reader, when presented with the results of a discourse analysis. The choice of category system and the inclusion of individual text fragments in these categories remain, of course, highly subjective. Each individual application of the schema to a discursive text will be compacted, first for each event, then as trajectories along the time axis, with the resulting concepts, objects, and subject positions always being a hybrid inscribed within groups of texts, and synthesized and formed by the interpretive effort of the analyst (Keller 2010, 222).

This interpretive effort of translation between two textual realities is guided by contextual and historical knowledge that the discursive-historical approach supplies. Justifications for these choices have to be highlighted and made as transparent as possible. These choices concern not only field selection, corpus assembly, sampling, and analysis, but also the way quality is ensured.

On Quality Control in Discourse Analysis

Challenges to Conventional Approaches to Quality Control

When using discourse analysis, or reading studies using such methods, one is quickly confronted with what is often seen as one a central problem of qualitative methods: the lack of a standardized way of assessing quality like quantitative studies do – or at least purport to do. How do quantitative studies solve this problem, and do these solutions apply here? Criteria that are often used to assess quality are *reliability*, *validity*, and *objectivity* (cf. Flick 2007, 15).

In general, *reliability* is concerned with what Stiles (1993, 601) calls "procedural trustworthiness": how trustworthy observations of data are. Reliability means that any given method can consistently measure what it is supposed to measure, that is, that it can be repeated, leading to the same results, by the same researcher or by someone else (cf. Lamnek & Krell 2016, 163). Of course, as Phillips and Hardy note, the idea that results can be repeatable is nonsensical in a discourse analytic context, since the goal here is "generating and exploring multiple – and different – readings of a situation" (Phillips & Hardy 2002, 80). Discourse analysis thrives on making different perspectives visible, that are subject to change depending on the context and reader.

Validity as "interpretive trustworthiness" refers to the trustworthiness of interpretations and conclusions drawn (Stiles 1993, 607). It is centered around the idea of whether a method captures the "real" world adequately. A variety of approaches were proposed to assess this, ranging from face validity via expert validity to construct validity (Lamnek & Krell 2016, 148f). Such a criterion goes against the foundational ontological and epistemological assumptions of discourse analysis. One of discourse analysis' core tenets is the rejection of the idea that there is a "real world" out there, of which language is only a mirror. Instead, the many ways that language creates the world in the first place are in focus (Phillips & Hardy 2002, 79). Of course, this understanding of the nature of language and reality radically questions other methodologies as well: "… when establishing knowledge about an aspect of reality, every research approach also makes specific assumptions about the nature of reality under investigation (ontology) and about the nature of knowledge (epistemology)" (Sandberg 2005, 47). From this vantage point, the ways that quantitative methods attempt to capture the "real world" are subject to processes of construction and power, and can be highly fertile grounds for discursive methods to elucidate how knowledge is created and legitimized.

The last criterion that is often used to assess quality is *objectivity*, which is based on the requirement that any given method is bias-free and intersubjective, and that it does not matter who is using said method. Once again, this criterion is not applicable to qualitative methods in general, and to discourse analysis specifically, partially because of the reasons already listed above, such as the development of different readings of texts. But most notably, the core of the idea that any research may ever be bias-free is questioned. As philosophers and sociologists of science such as Lakatos and Kuhn have shown, science is usually full of biases and "normal science" is a far cry from Popper's positivism; instead, they are often guided by overarching paradigms (cf. Stiles 1993, 613). Furthermore, scientists are in no way located at some Archimedean point:

> ... everyone is situated in a specific historical, cultural, and linguistic understanding of reality, which is internalized through upbringing, education, and work. The internalized understanding becomes to a large extent our framework for making sense of reality.
>
> (Sandberg 2005, 51)

Dismissing the effects of observer, context, and perspective would not only mean to ignore countless studies of the internal workings of science but also going against the core ideas of discourse analysis, and especially the context-focused historical approach developed here.

This brief discussion reveals why quality control in discourse analysis is worlds apart from how quantitative studies evaluate their methods, and thus why the three general criteria do not apply. As Flick (2007, 15ff) notes, there are three possible strategies for dealing with the shortcomings of conventional criteria while salvaging the possibility of assessing the quality of qualitative methods: One can either use the same quality criteria as quantitative studies, translate them, or develop new ones that are appropriate to the methods used (ibid., 18ff). Due to the fundamental ontological and epistemological differences outlined earlier that make it impossible to use the same criteria and near impossible to translate them, the third strategy is generally chosen in discursive work: the development of new criteria.

Four Criteria for Quality Control in Discourse Analysis

What criteria are commonly developed in the various approaches to discourse analysis, to solve the quality control problem? The approaches to evaluating discursive research vary in parallel to the multiple possibilities of operationalizing discourse depending on research interest and perspective. As there are few established norms for analyzing data and presenting the results of a discourse analysis (Phillips & Hardy 2002, 79), many different ideas exist for the problem at hand. Surveying the literature on discursive methods in general – and organizational discourse analysis in particular – criteria for quality control can be summarized in the following categories.

The first, *coherence*, is a criterion concerned with the interrelation of interpretations among themselves, as well as the larger context. According to Potter and Wetherell (1987, 170), the analytic claims made throughout a discourse analysis should give coherence to a body of discursive texts, both covering larger patterns and accounting for micro-sequences found within individual texts. The guiding question here would be how coherent conflicting interpretations are with the material (cf. Sandberg 2005, 55), and whether the interpretations developed in an analysis give a consistent account of the data at hand (Georgaca & Avdi 2011, 157). Several ways of ascertaining coherence are proposed, many of which are based upon testing researcher interpretations with actors in the field who produced the original accounts – an approach not possible in this case and in historical discourse analysis in general. Phillips and Hardy (2002, 80) underline another avenue for safeguarding coherence: by checking whether historical, contextual understandings are incorporated, and how the interpretations developed throughout an analysis fit with the events and trajectories of the field analyzed.

The next criterion, *rigor*, relates both to method and analysis, while also having an impact on the presentation of results. Rigor of method refers to sound methodical practice, starting with surprisingly simple tasks that are often circumvented when using a hazy concept like discourse: the development of clear definitions of concepts and their relations to each other (cf. Alvesson & Kärreman 2011a, 1195). What is discourse, and what does it do? How do concepts, objects, and subject positions relate to discourse, and each other? – and so on. When rigor of analysis is lacking, a discourse analysis may fall into the trap of under-analysis (cf. Burman 2003), being guided by pre-conceived or unequivocal interpretations. A discourse analyst should thus probe for ambiguity by giving "attention to inconsistency and diversity, analysing deviant cases in order to delimit the applicability of data, and providing richness of detail" (Georgaca & Avdi 2011, 157). Analytical rigor can be achieved by avoiding "armchair research" based on thin material (Alvesson & Kärreman 2011a, 1194), instead aiming for thicker descriptions and broad discursive coverage. Of course, alternative and often conflicting interpretations are hard to make visible, since they are necessary byproducts in the development of a larger, coherent interpretation, which is the one that ends up in the written account of the analysis. This is due to a text such as this one being the end-product of an analysis, and not a processual one.

The conflicting demands of rigorous interpretive work and clear presentation lead to the next criterion: *transparency*. Here, the goal is to make both the methodical approach to data as transparent as possible, while also reporting on results in a clear, accessible manner. With regard to methodical progression, "researchers must demonstrate how they have controlled and checked their interpretations throughout the research process: from formulating the research question, … obtaining data ..., analyzing the data obtained, and reporting the results" (Sandberg 2005, 59). The need for transparency poses challenges especially for the presentation of results which can be a daunting task when attempting to concentrate plenty of material in a few pages. The selection of excerpts in particular, arguably an important way of grounding an analysis (Georgaca & Avdi 2011, 157), can lead to difficult choices – most notably with an abundance of discursive texts as in this analysis (cf. Wood & Kroger 2000, 183f). Transparency in presentation goes even further, toward what Schwab-Trapp (2010, 184f) calls formulating and reflecting interpretation. In writing up the report, it needs to be made clear when extracts or summaries of texts are presented (formulating interpretation), when the analyst's (reflecting) interpretation generates grander connections between texts, and when the construction of concepts, objects, or subject positions is made visible in an interpretive motion (cf. Fairclough 1992, 198f; Stiles 1993, 605).

Making methodical progression and interpretive work visible is, of course, not a tool for feigning any forms of objectivity. In order to deal with the ways that reality is constructed through the process of interpreting it and in turn talking (or writing) about it, the last criterion that any discourse analysis needs to adhere to is *reflexivity* (Potter & Wetherell 1987, 182ff). This category is concerned with uncovering biases. Needless to say, researcher biases may creep in at many points: having a "favored" discourse (Grant & Marshak 2011, 225) or assuming in advance what elements of social reality are at play and how they relate to each other (Alvesson & Kärreman 2011a, 1995). The goal of reflexivity in discourse analysis is to develop interpretive awareness, which means to "acknowledge and explicitly deal with our subjectivity through the research process instead of overlooking it" (Sandberg 2005, 59), that is, to be aware of one's own subjectivity throughout the analytical work. Of course, there is no obvious path toward reflexivity. Instead, it can be understood as an iterative process in which both coherence and rigor play a large part and a researcher continuously checks their interpretation of discursive material, how the interpretation was developed, and how their "perspectival subjectivity" (Sandberg 2005, 59) may have guided the interpretation. Phillips and Hardy (2002, 85, Table 5.1) offer guiding questions for assessing reflexivity throughout the research process that emphasize the constructive nature of language both in discursive text and in the researcher's report. Stiles (1993, 613ff) takes a particularly optimistic position in regard to reflexivity, noting how, despite biases, expectations are repeatedly disconfirmed, and close engagements with research material by means of thick descriptions and (contextual) triangulation may offer a new response to the idea that the possibility of biases invalidates any findings whatsoever. Nevertheless, developing a reflexive approach can be seen as one of the key challenges of discourse analysis.[5]

Notes

1 Even though the organizational field cuts across delineations normally drawn such as "academia" or "industry", much of the corpus material can be regarded as academic discourse. Today's common practices for this genre (Hyland 2011) bear little similarity to those of 1870–1930, yet studies on the genre's genesis (Bazerman 1988, esp. 80ff) aid in adding another layer of context to the analysis, especially in understanding how the genre's conventions were shaped by the structural developments of science, and in turn influenced its shape through the mediation of legitimate speakers and contributions. Similar works exist for popular science writing, helping to place texts contributed by popular science journals (cf. Mott 1957a; LaFollette 1990).

2 To name but one example: Searching for the term "R&D laboratory" would hardly have yielded any results, since the early laboratories were not named that – the concept of "development" only occurred several decades later (see above, also Godin & Lane 2011). Instead, informed by the detailed accounts on the early laboratories recounted in Chapter 3, the proper terms for such laboratory settings used at the time were extracted.

3 105 texts had either no author or could not be attributed without error.

4 As Howard Becker wrote in *What about Mozart?*, a key ability in qualitative analysis is to know when to "finish the damn thing, with all its flaws" (Becker 2014, 165).

5 In discourse analytic literature discussing quality control, one often finds the additional criterion of "usefulness" or "fruitfulness" (Potter & Wetherell 1987, 171; Phillips & Hardy 2002, 80; Flick 2007, 21; Georgaca & Avdi 2011, 157), as in the power of an analysis to reveal new insights about the world or to make theoretical contributions. Since this facet is seen as one of the key elements of any science no matter the methodical approach (Whitley 2000), it is not listed as a possible criterion for quality here.

6 Analyzing the Discourse

After the difficult decisions involved in conducting a discourse analysis, the analyst is confronted with the next problem: how to present the results. Akin to operationalization and corpus generation, there exist few established norms or standards for writing up a discourse analysis (Phillips & Hardy 2002, 78f). The analyst needs to negotiate multiple pressure points: illustrating with examples from texts while not going overboard and simply narrating the content of the texts analyzed, articulating clearly where texts end and analytical interpretation begins, as well as alternating focus on discursive parts but not losing the big picture of the discourse. In order to negotiate these tensions, several key ideas shall be followed. First, writing up the results of a discourse analytic study is itself, of course, a discursive activity (Wood & Kroger 2000, 179), starting with the choice and framing of research and going all the way to the selection of textual examples that find their way into the study. Discursive effects cannot necessarily be steered clear of, especially since they originate from the interaction between texts, as well as their production, dissemination, and reception – activities that the writer has little to no influence on. Thus, it is important to be as transparent as possible, clearly illustrating claims by way of the analytical steps that led to said claim (ibid., 182), as well as making the additional and contextual information affecting interpretations and decisions as explicit as possible.

Second, one needs to be aware of when the descriptions of texts and their respective interpretations merge. This is due to analysis being an iterative process, with interpretations forming from descriptions of texts that are then re-interpreted in light of discursive features that were not clear before (Fairclough 1992, 231). Furthermore, in these interpretations there is a "constant alternation of focus from the particularity of the discourse sample, to the type(s) of discourse which it draws upon" (ibid.). These interlocking practices make it difficult to trace the way that an analysis was performed and to present it in a comprehensible way to an outside reader, especially since simply listing the temporal sequence of the analysis may not clearly show how some interpretations came about.

Third, in response to these problems, Schwab-Trapp (2010, 184f) proposes two steps for the precise and comprehensible presentation of the results of a discourse analysis. In the first step, which he calls "formulating interpretation", the manifest contents of the texts shall be described, as close to the textual material as possible. The second step, "reflecting interpretation", will go beyond the text and reconstruct how the textual elements relate to one another, how they are used, and how they relate to other texts and the broader discourse. In this step, the way that concepts are materialized in objects will be shown, as well as what roles the actors who are in legitimized subject positions play in the process. As Wood and Kroger note, the final report represents a concise and selective presentation of the analytic work (Wood & Kroger 2000, 186) – by design, some texts and some findings will have to be left out. Instead, "key texts" that illustrate certain arguments or features will be discussed, and their relation to other texts, as supporting or dissenting voices, will be made clear. The reasoning for designating a text as "key" will have to be made transparent as well. Thus, the presentation of results will follow Schwab-Trapp's distinction, first in describing the key texts, then in opening up the textual material to interpretation as discursive fragments.

Before we delve into the many ways that the industrial research laboratory was made and remade as a discursive product, some general remarks on the discourse as a whole are in order. First of all, the texts representing discursive fragments – parts of a larger discourse that the analysis attempted to reconstruct – turned out to be surprisingly homogeneous. There were no warring factions of "pure" and "applied" scientists contesting each other and denying the others' claims of being scientists or designating the industrial laboratory as a travesty. There were no industrialists denouncing scientists' claims of scientific efficacy. Also, there were no groups utilizing distinctly contrasting argumentations. Instead, the construction of concepts, objects, and subject positions was rather homogenous across texts, with little variation within events: Changes mainly happened over time, and less so in relation to the position of the speaker. In turn, the lack of antagonistic groups of speakers was similar to the lack of antagonistic concepts. In the discourse surrounding industrial laboratories, it appears that there were hardly any competing concepts of who the chemist is or where their place of work may be. Thus, the story of this analysis is not the mapping of how and why one concept or subject position was victorious and another vanished from the discourse, but rather how these dominant ideas came to be constructed, what resources were used in their argumentation and legitimation, and – most importantly – how they changed over time, in an organizational field that spans sixty years, multitudes of institutional and organizational change, and one world war. One could hypothesize that the discursive homogeneity is due to the choice of speakers, and for more contrarian voices to be heard, other groups – such as inventors, businessmen, and others – need to be included in an extended discourse analysis. The choice was made to assess constructions of science, the scientist, etc., by "scientific" speakers in the broadest sense, covering the outlets of scientific societies and science-adjacent professional associations, as well as popular science journals. This decision further tailored the discourse covered based on the idea that in order to find out whether any kind of cultural shift happened at all, it would be necessary to find out how science (etc.) was constructed by scientific speakers themselves as the primary carrier group of the discourse on science, which was then carried outward toward industrial organizations, political appointments, or the public at large. In this way, the analysis undertaken here makes no claim of covering the entire discursive reality of the organizational field of corporate R&D; instead, the methodical path taken, especially in drawing the boundaries of the organizational field and the choice of speakers for corpus assembly, was guided by the decision to focus on the cultural shift toward science as proposed by prior historical scholarship (Tobey 1971, 12; Hughes 1989, 138ff; Friedel 1992; Hintz 2011).

Next, not all categories proved to be as relevant as expected. Fairclough put it quite succinctly when he stated that "in any particular analysis some of the categories are likely to be more relevant and useful than others, and analysts are likely to want to focus upon a small number of them" (Fairclough 1992, 231f). As it turned out, the notions of genre and intertextuality played only minor roles in the analyzed corpus. In regard to genre, texts were not subjected to a full-blown genre analysis, rather I looked for structural similarities or peculiarities. As expected, an address differed from an article and from a letter to the editor in composition and also often in tone or style. Also, few differences in genres between groups of speakers could be identified, possibly due to the journal-based nature of the corpus. The minor part that intertextuality played may come as a surprise, with a good subset of the texts being articles published in scientific journals, where citing would be part of the practice of scientific writing (cf. Fairclough 1992, 128). Yet many of these texts could better be described as addresses or essays, negating the imperative of attributing sources. So, in the first place, the existence of manifest references was noted. When those texts that used intertextual references were analyzed as a group for patterns – such as frequent references to certain texts (both intra- and extracorporeal), books, laws, sayings, etc. – hardly any emerged. Instead, a large group of references could be collected – ranging from scientific works and newspaper articles to books such as the Bible – that were indicative of the absence of accepted, central texts that could be used for legitimation by mere reference. More revealing than explicit intertextuality were the less manifest forms, such as recurrent phrasings or turns of expression associated with subject positions, for example, the chemist or the banker.

Overall, the lack of central texts supplying argumentations and ideas that served as legitimating resources may be due to the heterogeneous nature of the organizational field, which brought organizations together whose different cultural backgrounds supply different discursive standards.[1] In fact, clear patterns did not emerge even when the corpus was analyzed as distinct groups of speakers with possibly different discursive backgrounds (that is popular, chemical, industrial, and governmental). Thus, intertextuality played only a small role in the materialization of the laboratory and the legitimation of the chemist as an important actor in the corporation.

What proved more interesting was what is generally called "interdiscursivity" (Fairclough 1992, 124; Phillips et al. 2004, 644). Interdiscursivity means the drawing upon other discourses as support or demarcation, importing related concepts and their "particular way of constructing a subject-matter" (Fairclough 1992, 128). Unsurprisingly, the organizational field analyzed here played host to a multitude of discourses, some pertaining to field-level issues, with others being more society-spanning. Such discourses influence each other and can be used as resources for bolstering one's own discursive constructions (Phillips et al. 2004, 644). This relies on the central tenet that discourses are linked to each other, hardly ever occurring alone and often in groups. These other discourses – that texts from the discourse analyzed here, on the emergence of industrial research laboratories, linked to – were mainly found through the topical patterns noted in the schema of analysis. As briefly mentioned earlier, when circumscribing topics of each individual text, it soon became apparent that for single (or adjacent) events one or sometimes several topics were prevalent, which is why the analysis will start with an introduction of the nine other discourses that were of concern in the field during the time of analysis, outlining their central concerns and arguments as well as broader, societal framing. How concepts, objects, and subject positions were constructed and had to be translated from event to event can only be decoded if these dominant topics, as imported fragments of other discourses, are properly understood.

The main findings of the discourse analysis are summarized in Table 6.1. For each event, other discourses emerged, whereas concepts, objects, and subject positions were translated by a variety of speakers, thereby building bridges toward said other discourses, reacting to events in the field, and arguing for or against ideas, all while constructing and re-constructing the idea of the industrial research laboratory in new and unique ways.

Interdiscursivity

The nine discourses that will now be described in detail emerged through the analysis of the texts and were initially not actively searched for. That

Table 6.1 Overview of Results

Event	*Discourses*	*Concepts*	*Objects*	*Subject Positions*
1895 Professionalization of Chemistry	Chemical Education Specialization	Science: • quest for truth • superiority over nature • scientific method • progress, material wealth & welfare through industrial applicationsChemistry: • similar attributes as → Science • constructed as the fundamental science • transformation of matter, closeness to nature • utility in practical, industrial applications	Laboratory: • explicit association with attributes of → Science • scientific method only possible in Laboratory • practices in laboratory: teaching and research • staffed by professional scientists	Chemist: • man of science • university-educated • applicator of the scientific method, practitioner of → Research • similar relation to truth, nature as → Chemistry • teacher & researcher • personal qualities: honesty, truthfulness, accuracy, … • forefathers of chemistry characterized as special, geniuses
1900 GE Lab	Chemical Education		Laboratory: • the place of employment for → Chemists	Chemist: • constructed as pioneers, surveyorsChemical Engineer: • first mentions of the term

(*Continued*)

Table 6.1 (Continued)

Event	*Discourses*	*Concepts*	*Objects*	*Subject Positions*
1902 DuPont Labs	Chemical Education Specialization Research Landscape	Science: • becomes synonymous with → ResearchResearch: • practice of "doing science" • organized, slow, methodical • similar relationship to truth, nature as *Science*	Laboratory: • connected to → Research Landscape: calls for laboratory endowments • emphasis on industrial applicability of laboratory findings	Chemical Engineer: • discussion whether Chemical Engineers are chemists or engineers • need for → Specialization
1908 ACS Division IEC	Specialization Conservation Movement Efficiency	Chemistry: • for improved → Efficiency & → Conservation • vs. rule-of-thumb methods • increased distinction: Analytical ↔ Research Chemistry	Laboratory: • for → Efficiency • increased distinction of laboratory types: analytical, testing, research, …	Chemist: • increasing distinction of types: Analytical Chemist, Research ChemistIndustrial Chemist: • emergence of the term • seen as "technical" chemistChemical Engineer: • discussion settled: chemical engineer as chemist with an engineering side

				• similar attributes as the → Chemist • applies → Chemistry in an industrial context • prime actor in → Conservation
1912 Kodak Lab	Conservation Movement Efficiency Research Landscape Legislation	Industrial Research: • similar attributes as → Research • foundation for industrial progress • linked to → Applied ChemistryChemistry: • increased discussion of Pure, Applied Chemistry	Laboratory: • seen as a necessity in industrial organizations • two main types: Analytical, Research, relates to → Analytical Chemist, Research Chemist	Chemist: • → Research as a group activity • construction of hierarchy: Research Chemist superior to Analytical Chemist • no clear localization of Analytical, Research Chemist • in opposition to foremen (rule-of-thumb methods, empiricism) lacking → Efficiency
1917 WWI	Research Landscape Preparedness Legislation Popularization	Science: • progress of Applied Science due to advances in Pure Science • Pure Science: slow, unplanned,	Laboratory: • for → Preparedness • German example: military power through laboratory research • increased discussion of proper setup &	Chemist: • lack of appreciation by managers, need for → Popularization • need to educate the public on the values

(Continued)

Table 6.1 (Continued)

Event	*Discourses*	*Concepts*	*Objects*	*Subject Positions*
		gambleIndustrial Research: • for → Preparedness • German example • need to sell value of IR to managers (→ Popularization)Chemistry: • for → Preparedness • German example of industrial & educational organization	equipment of industrial laboratories	of → Chemistry (→ Popularization)
1922 DIR	Efficiency Research Landscape Legislation Popularization Postwar Normalcy Chemical Education	Chemistry: • need for → Popularization, → Legislation		Chemist: • call for → Popularization of the chemist's services and achievements • → Legislation to safeguard place of employment

Links to specific discourses, concepts, objects, or subject positions are denoted with "→".

is to say that, prior to the analysis, it was not evident, say, what role the Conservation Movement would play for the events of 1908 and 1912, but its importance was revealed since many of the texts of these two events referred to or at least mentioned the movement. Figure 6.1 shows a rough overview of these interdiscursive trends, especially their starts, peaks, and ends measured as the percentages of articles analyzed for that year that discussed the discourse in any way.

As one can see, there exists a distinct progression of discourses, with these discussions disappearing after one or two events and new ones occurring. But there is also overlap and coupling, such as with the case of the discussions of the Conservation Movement and Efficiency, or Science Popularization and the return to Postwar Normalcy. The only other discourse that was discussed across all of the years – sometimes not as vigorously as others, but playing a part nonetheless – was chemical education – how to properly teach chemists and equip them for their academic or industrial future.

Specialization

The discourse on Specialization is concerned with the increasing differentiation of academic chemistry, both disciplinary as well as organizational, and was an area of discussion primarily for the events of 1902 and 1908. Already lamented around 1870 (Bruce 1987, 347), disciplinary specialization saw old specialties wane and new ones dominate: Prior to 1900, analytical, inorganic, and organic chemistry were the major subsets of chemistry, whereas after 1900 organic and physical chemistry started to dominate (Rothenberg 2001, 116; cf. Ihde 1964, 531), with the idea that one professor could cover the whole field having disappeared long before (Beardsley 1964, 49). As the history of the organizational field of chemistry has shown, organizational specialization occurred not only with the founding of new organizations such as the AIChE, AES, and the American Society of Biological Chemists but also through the divisional structure of ACS.

Interestingly, the discussion on specialization in the texts analyzed proceeds along similar lines and is one of the few topics covered where strong oppositions could be distinguished. On the one hand, there was a strongly optimistic strand of discussion that saw specialization as a fundamental prerequisite for scientific progress (Wiley 1902). Specialization in science was often likened to a concept from the industrial world: the division of labor. These discussions frequently occurred in the context of the establishment of new organizations, namely the AES and the AIChE, and organizational specialization was seen as running parallel to intellectual specialization:

> Differentiation and specialization are the watchword, now, of all progress, – industrial, scientific, philosophical. The day is past, we all

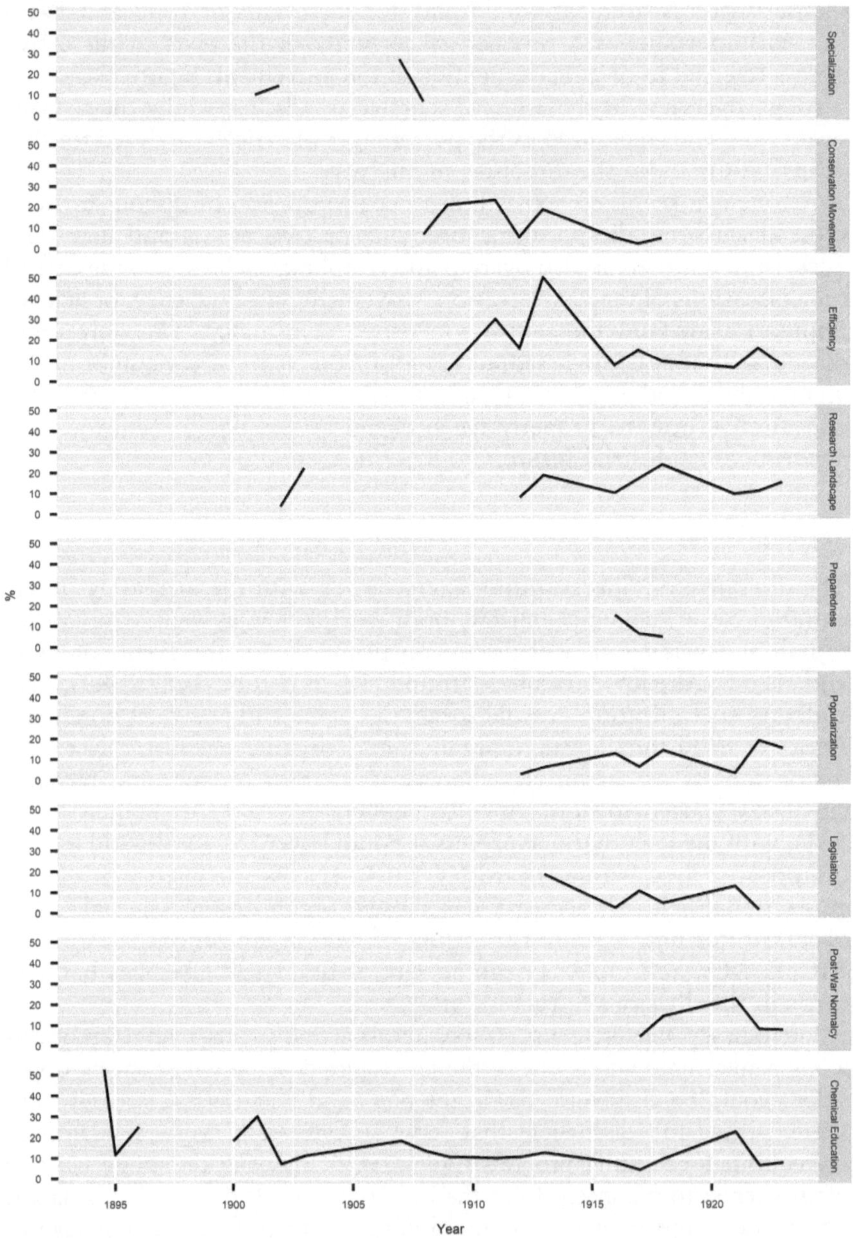

Figure 6.1 Percentage of Texts Referring to Certain Discourses.

> acknowledge, when one man, even be he Newton, can know all that is to be known; the day is also past when one scientific society can cover satisfactorily the whole field of scientific research. Even more than this, the day is passing when any one society can even cover satisfactorily the whole field of any one science, such as physics or chemistry or medicine.
>
> (Richards 1902, 1)

In this way, the designated "age of specialization" ("Why Not 'The American Society of Chemical Engineers'" 1907, 227) and all its positive connotations was apparently used as a rhetorical tool justifying the need for organizational plurality, especially with the establishment of organizations outside of the ACS divisional structure (cf. Hillebrand 1907).

On the other hand, specialization was rendered as negative or even dangerous to the discipline of chemistry when put in the context of education. The dissociation of chemistry into its various subdisciplines was seen as the cause for highly specialized chemists losing touch with other branches and failing to engage with or even understand each other (Noyes 1908). Thus, it was stipulated that specialization should not interfere with education, which was to be kept as broad as possible (Wiley 1902; Richardson 1908). If the university only produced narrow specialists it was seen as having failed at reaching its ideal: the production of "broad, scholarly minds" (McMurrich 1907, 646). Ultimately, specialization for employment in industry should come after education in the fundamentals, which was to be kept as broad and inclusive as possible. Discussion on the advantages and dangers of specialization wore off after the 1907–1908 event tranche. This may be due to other topics clouding these discussions, as well as the organization of (academic) chemistry in scientific and related societies reaching a certain saturation at the end of the decade, with most of the major societies having been established, especially in regard to the organizational home of the chemical engineer. Furthermore, the ACS established its divisional structure by 1908, with the intensity of the discourse prior to this event, as well as its diminution serving as an illustration of how the reality of the organizational field is reflected and reproduced within the discourse.

Conservation Movement

The Conservation Movement was a relevant adjacent discourse for the events of 1907–1909 and 1911–1913, with the discussion reaching its apex during the second period. It can be seen as a prime example for other discourses finding their way into the discourse on industrial research laboratories (Phillips et al. 2004, 644). While other discourses, such as Specialization or the Research Landscape, resound in events and processes on the level of the organizational field such that reflection in discursive texts could be anticipated to a certain extent, the Conservation Movement can be regarded as happening outside the confines of the organizational field and as a movement on the larger societal plane, supplying additional discourses that may find their way into the specific discourse we are interested in. To give some brief context, the Conservation Movement, anchored firmly in the Progressive Era, grew out of the experiences of economic growth in the late nineteenth century (cf. Wiebe 1967, 164ff), which resulted in crime, overcrowding, poverty, pollution, and uncontrolled industrial development (Taylor 2016, 1ff). Conservation signified "a utilitarian view of natural resources: that is, they should be developed and used for the current generation" (Taylor 2016, 27). The Roosevelt administration came to be especially linked to the movement, with Theodore Roosevelt himself being a strong advocate for its aims (cf. Hays 1959, 14; Taylor 2016, 73ff, 387). The Governors' Conservation Conference of 1908 put conservation issues in the limelight and lead to the establishment of the National Conservation Commission to undertake an inventory of all natural resources (Hays 1959, 128f, 140). Even though the leaders of the movement managed to cultivate a certain pro-conservation public sentiment (involving strong support by the major engineering societies [Hays 1959, 123f; Daniels 1971, 302]), the movement's powers were soon drained with the beginning of the Taft administration, as a skeptical President Taft vouched for policy changes (Hays 1959, 147ff). By the 1910s, conservation had become a highly elastic term, meaning "vastly different things to different people" in practice (ibid., 175), and the time of an organized and powerful movement was seemingly over.

The movement was discursively connected to the discipline of chemical engineering, and especially its subject position in the chemical engineer. With direct reference to the National Conservation Commission, conservation as the proper development and application of resources was claimed to be "'almost a good definition of chemical engineering'" (McKenna quoted in Roeber 1908, 309). The chemical engineer was seen as playing a pivotal role in conservation efforts to curtail the wasteful use of natural resources (Richardson 1908; Bogert 1913) and in waste management or reduction (Bailey 1911). Furthermore, the chemical engineer could aid in conservation efforts applied to industry, such as broadening its scope and pushing for stability and permanence (Sadtler 1909, 106f). Next to chemical engineering, electrochemistry was often mentioned in the same breath in regard to the Conservation Movement. Heeding the Governors' Conference's call, the field's scientific knowledge was to be used in both preserving the United States national assets, and in an energetic yet conservative utilization of existing resources. Exploring and developing the field would be the duty of the electrochemist, who would "undoubtedly" play an important role in conservation. Electrochemistry's contribution to conservation revolved particularly around the generation of hydroelectric power at Niagara Falls, a topic of contention for the movement, with regard to the question whether conservation was to subordinate aesthetics to utility (cf. Hays 1959, 127). In the discursive texts analyzed here, the taming of Niagara was celebrated as one of the brightest chapters in the history of the American industrial revolution, with the electrochemist having shown the public how Niagara Falls could only be perceived as "beautiful" if its humongous reservoir of power did not go to waste (Roeber 1911a, 1). Conservation's elasticity as a concept is illustrated when turning to some texts in the period of 1911–1913, when political support for the movement waned. Discussions stopped focusing exclusively on natural resources and the (electro-)chemist's or chemical engineer's part in the movement and broadened toward conservation itself as a topic for chemical research ("The Conservation of Research" 1911; Cottrell 1912), as well as the application of scientific management as the principles of conservation to production in chemical industries or even universities (Gillett 1913, 594, 599). In the increased spread of ideas regarding the proper application of scientific management a central element of the Conservation Movement's tenets is illustrated: the notion of efficiency, which created such a resonance in discursive texts that it will be discussed separately.

Efficiency

The notion of efficiency was most certainly the part of the Conservation Movement that resounded the strongest in the discourse analyzed. Playing a part in the last four-event tranches, discussion of efficiency in industry, academia, and government peaked around 1913. The Progressive Era "gave rise to an efficiency craze" (Haber 1964, ix; cf. Galambos 1979, 274f). Lying at the heart of the idea of the conservation, the "gospel of efficient planning" concerned not only the efficient use of (natural) resources but also all areas of life (Hays 1959, 124). Nothing marked the progressive movement more than this interest in efficiency, with which science came to be identified – the application of the scientific method to life being believed to yield greater efficiency (Daniels 1971, 308). Of major concern to the discourse analyzed here were two readings of the term in particular: efficiency as mechanical efficiency, or the energy input and output of machines; and the efficiency of the business enterprise, as input and output of dollars (Haber 1964, ixf). Corporate receptivity to the craze fits well with Fligstein's assertion that in the manufacturing perspective – the dominant conception of control during these years – business stability was sought through the internal control of the production process (Fligstein 1990, 16), with a more efficient process evidently leading to higher profits (cf. Barley & Kunda 1992). The application of efficient measures to business and other areas became associated to a large extent with Taylor's Scientific Management, which set out to apply scientific methods to business problems (Haber 1964, 18ff), underscoring how the discourse on the credibility and efficacy of science lends legitimacy to these linkages of efficiency through science (Daniels 1971, 303; LaFollette 1990, 9).

Efficiency was depicted as the means to ameliorate conditions in industry (Duncan 1909), the way to create greater prosperity (Roeber 1911a; Little 1913) and the answer to the waning of "[t]he days of large profits, cheap raw materials and labor" (Booth 1912, 196). The orderly use of talents and especially research staff (Ferguson 1912) was seen as the path to increased efficiency, with the chemist built up as the proper actor who could apply scientific insights to industrial processes – a fact that was lamented as being overlooked by manufacturers (Skinner 1911). But the concrete methods to improve efficiency were characterized not only by subject positions but also by the concept of research and its materialization. Efficiency could be achieved through research as the "catalysis of raw materials by brains" (Little 1913, 644) and the use of chemistry (Little 1911). In fact, some authors went so far as to describe the industrial research laboratory as "an extension of the principle of maximum efficiency" (Whitney 1911, 429; cf. also "The Industrial Corporation and the Inventor" 1911), which indicated the spread of the concept of laboratories in industry, and possibly further underscores how nascent and emerging entities are legitimized by translating them within the given frame, in this case: a discourse that was crazy about efficiency (see later). As outlined earlier, the quest for efficiency did not cease at the boundaries of the business world. The concept was applied to the training of chemists and engineers, and questions were asked regarding whether or not contemporary training was efficient enough for employment in industry (Burgess 1911; Withrow 1911), with some calling for an "atmosphere" of efficiency in laboratory instruction (Whitaker 1911). Calls such as this were often informed by assertions that the university chemist was not efficiency-minded (Bancroft 1912), and therefore limited in their employability in the manufacturing sphere that was only concerned with profits and efficiency (Grosh 1913). Beyond teaching, proper equipment was demanded to increase research efficiency (Roeber 1911b). With the beginning of WWI, the way that efficiency was linked to concepts, objects and subject positions shifted, further illustrating the fluidity of discursive concepts, as well as the approach of using event-tranches as a reasonable approach to map changes of meaning. The war was described as an international striving for efficiency, with the German military-industrial war machine being dreadfully efficient (Noyes 1917). The key to success was depicted, unsurprisingly, as efficiency through science (Bogert 1917b, 1010; cf. Withrow 1917). Lastly, the craze also produced demands for field-level cooperation in research: "[C]loser coöperation between our universities and our industries will go far to assure increased efficiency to both" (Bogert 1913, 762).

Research Landscape

Discussion of the Research Landscape occurred mostly between roughly 1903–1907 and 1912–1922. The primary concern of this discourse was support for chemical research and its institutional location. In the early period, proposals for possible and proper support of research were communicated and discussed – mainly framed as research happening at universities, where it would have to scope out a place for itself next to teaching duties (Chamberlin et al. 1903). The Carnegie Institution emerged as a highly visible actor, with listings of grants made repeatedly reported on the pages of Science and other journals ("Grants Made by the Carnegie Institution" 1903).

While the Carnegie Institution remained visible and heralded in the second period, 1912–22 ("Ten Years of the Carnegie Institution" 1912), other ideas came to the fore, such as the Research Corporation (Cottrell 1912; "The Research Corporation" 1917; Robertson 1917), where profits from patents were utilized to fund research, and a system of industrial fellowships (Duncan 1912), where corporations (such as Du Pont) supplied funds for selected research projects at universities. A subtle shift in direction and emphasis marks the second period, especially with the entrance of the United States into WWI. Whereas at prior events the discussion centered around the question of establishing systems for research funding – that is, how to fund research at all – now the discussion focused on what the *best* way to fund research would be, how to facilitate cooperation between universities and corporations, and how to coordinate laboratories to eliminate duplications of effort, seen as waste of researchers' manpower and time. The difference becomes especially clear once the question of independent laboratories occurs (Sharp 1917), as the high costs of establishing an in-house research laboratory became widely known – these were costs that smaller manufacturers could not afford, hence the need to collectively shoulder the load (Carty 1916; Weidlein 1922).

Proposed solutions were the services of independent consulting laboratories or the establishment of cooperative laboratories by trade associations or groups of manufacturers. This strand of discussion intensified, with the end of the war in sight and speakers setting their sights on the postwar configuration of the American research landscape (Bancroft 1918; Wardenburg 1922; de Long 1922). In this second period, the relationship between universities and corporations and the proper place of research was also increasingly broached. A clear duty of industry to "give back" to academia by funding university research (through various proposed measures such as endowments, fellowships, etc.) was evoked time and time again (Bancroft 1918). The displacement within this topic can be seen as indicative of an evolving research landscape, which is further supported by field-level contextual data. At early events, prior to 1912, relations between universities and corporations were still nascent, with the university still in the process of balancing its functions of teaching and research and the industrial research laboratory not yet a well-known successful entity. During and after WWI, places of research became more clear-cut, and it became taken for granted that research was done not only at universities but also within corporations. The discussion on the proper cooperation and coordination of research efforts nicely illustrates how the industrial R&D laboratory had become a known organizational element within the discourse, so that even those without the means to afford a laboratory needed to find ways to reap the benefits of research, further underlining how the concept of research gained traction as the foundation of industrial progress over the years.

Preparedness

The discourse on Preparedness only touched one event, the tranche of 1916–1918, but it reframed many of the established conceptualizations of science, chemistry, and research. With the war already waging in Europe and the United States' entry looming on the horizon, discussions on how to properly prepare – not only militarily but also industrially – for such an event reached heightened intensity. Many different ideas were espoused, to counter the state of "deplorable unpreparedness" (Carty 1916, 511), and research was positioned as the foundation of proper preparedness and "America's bulwark of defense" (Parr 1918, 419; cf. Bacon 1916). This is especially interesting since the horrors of the war, trench warfare and chemical weaponry, were seen as products of the (primarily German) sciences and industries (Carty 1916, 511). Nonetheless, research was perceived as a national duty (Whitney 1916), as the nation's military power was based directly upon the execution of scientific research and its application in the industries (Bogert 1917a, 1917b; Withrow 1917). Chemistry was portrayed as the center of these activities, with authors enumerating the ways in which chemistry touched not only agriculture and the military but also daily life (Raiford 1917, 493). Research and the implementation of preparedness-related requirements were put at the feet of scientifically trained men, who were seen as natural leaders, and their training came under increased scrutiny (Maclaurin et al. 1916; "Where are the Leaders?" 1918). The scientist and especially the chemist were also discussed in ways connecting to both the topics of the Research Landscape and Popularization. The Research Landscape was included in the question of the proper place of the chemist (and by extension the scientifically trained man) in the war effort, with several authors lamenting the drafting of highly qualified chemists into frontline duty, thereby depleting research laboratories at universities, corporations, and federal institutions. Instead, they saw the chemist in the reserve doing research not only to supply increased industrial yields and wartime goods but also to solve problems under the direction of the Army and Navy (Bogert 1917a), in this way fulfilling their "patriotic duty" (Stieglitz & Parsons 1917, 730). As explained with the history of chemistry and its relation to politics and the war, the NRC was founded in an effort to better coordinate research work between academia, industry, and the military, in addition to helping chemists find their "proper place" within industrial and military organizations geared for wartime. The reason for its organization and subsequent actions is mirrored in discursive texts on Preparedness. With regard to Popularization, the preparedness campaign and successive transformations of the organizational field were seen as indicators of ruptures in the perception of chemists and scientists in general, as well as a public reckoning of the chemical foundations of all societal life (Ames 1918). This brings me to the next adjacent discourse: Popularization.

Popularization

The popularization of science was discussed during two events, 1916–1918 and 1921–1923. This topic is especially interesting due to the insights it opens up regarding construction and change in subject positions as well as concepts. With regard to subject positions, ideas about who the chemist is play a central role. Prior to entry into WWI, during the height of the preparedness campaign the public perception of chemists – or rather, the perception discursive speakers had of the public view – was critiqued repeatedly. Even though the preceding years were seen as an upswing regarding popular knowledge of what chemistry is and what a chemist does, "[t]he popular mind has evidently not yet progressed beyond the age of the alchemist" (Matthews 1916, 1148). This was perceived as weighing especially heavy on the opportunities for chemists in industry since bankers were not prepared to invest into chemical businesses due to a lack of familiarity ("The Chemist and the Banker" 1916). Similarly, the businessman was unready to change processes and trust the chemist on the shop floor (Maclaurin et al. 1916, 65ff) due to a lacking of appreciation for science: "The ordinary business man of to-day very naturally thinks that his business ability is more valuable than the technical ability of the chemist" (Thompson 1917a, 81). A remedy was supposedly found in the continued efforts of popularization, of which Duncan's books and the activities of the USDA and Department of Commerce were especially lauded (ibid., 80). Moreover, the need to transform public perception was strongly associated with the attribution of blame for the dyestuffs crisis, originally laid at the feet of the American chemists, who saw themselves wrongly accused and in turn emphasized the contribution of chemists to turning the situation around (Clark 1917a; Herty 1917; Parr 1918). Many speakers praised the ability of American chemistry to build up the knowledge base for dye manufacturing so quickly, and shift the blame toward industrialists unwilling to invest in dyestuffs prior to the crisis (Baekeland 1917; Bruce 1917). Thus, popularization materialized in various ways – through books, newspaper articles, public exhibits, and speaking tours – was regarded as a means for increasing appreciation of the chemist's subject position in the public eye, as well as popular concepts of science and chemistry, after the nation's awakening to the needs of a strong national chemical industry. As the discussions surrounding Preparedness have shown, science and research were constructed as central to the war effort, notions that were echoed in the efforts of popularization. After the war, discursive texts show a high willingness to exploit the war's "momentum" with regard to public knowledge of the powers of science and chemistry (Campbell 1921; Slosson 1922), and to make it known to everyone how exactly the war was won and how integral the organized use of science was not only to successful military action but also to building a strong, national chemical industry that would be able to compete internationally and supply American markets independently of foreign suppliers. This is also evidenced by the organizational field, with the ACS News Service and the Science Service gearing up, and the NRC as well as the Chemical Foundation supplying their own brand of "chemical boosterism", especially in efforts to influence legislative efforts.

Legislation

Legislation was discussed during three events: 1911–1913, 1916–1918, and 1921–1923. Contributions gained intensity especially during and after the war, which is why this topic is intrinsically linked to others, that is, Preparedness, Popularization, and Postwar Normalcy. The discussion can be split into two strands: In one, legislation was seen as a protective mechanism for the American chemical industries, while in the other, legislation was expected to aid in the consolidation of the professional status of the chemist. The dyestuffs shortage was frequently invoked as a cautionary tale of what can happen if American industries remained dependent on foreign powers: the war showcased the need for industrial independence. The re-awakening of the German industrial powerhouse was painted as the "greatest fear" of American chemical manufacturers (Noyes 1917, 6). Celebrating the research accomplishments of American chemists who reacted to the dye crisis and achieved successes in the war, the onus was now put toward legislators to safeguard what was attained (cf. Herty 1917; Baekeland 1917).

Indeed, it was stressed that while the nascent dyestuffs industry in the United States was a result of chemical research in academic and industrial laboratories, chemists could not move the industry into a permanent and independent state on their own. Instead, "independence is altogether a question of capital, not of science" (Stieglitz 1917a, 2102) – capital was seen as a vital ingredient, concurrent to discursive reasoning in shifting the blame for the dye crisis, where it was not the lack of research, but the lack of will to invest that was lamented. To assure investors of the future prospects of American chemical industries, "wise legislation by tariff and patent laws will insure to capital a return sufficiently attractive and stable" (ibid.). Utilizing connections built up during the war, the ACS was seen as being in the ideal position to influence such legislative efforts through advisory committees and other cooperative activities (Stieglitz 1917b, 1006), illustrating the politicization of chemistry – not only did individuals get involved but also big, central organizations in the field such as ACS and the AIChE.

The other strand of the discussion revolved around chemists as a group and their professional standing. While the dye crisis and the war were seen as elevating the chemist not only in the public eye but also in those of industrialists, certain speakers noted that the profession was still not well-defined or established. References were regularly made to other learned professions – medicine and law – and the advantages that their legislatively mediated existence yielded (Bacon 1917a, 800f; Stieglitz 1917a, 2110; Seidell 1922). Thus, recognition by legislation would not only protect the public from chemical malpractice but also further increase the chemist's public standing (McCormick 1921). Some considered the major chemical societies able to regulate the professional requirements made of chemists to guard against the dangers of "impostors and unqualified consultants" (Bacon 1917a, 800), yet the issues of professional cohesion, perception, and legislative efforts for the stipulation of standards came up increasingly surrounding the establishment of the AIC as an economic organization for chemists distinct from scientific or technical societies (Parmelee 1921a; McCormick 1921; "American Institute of Chemistry" 1922; Eisenschiml 1922b). In this way, the economic struggles of chemists outlined in the organizational field above are reflected in the discourse, as questions of legislation for professional status and the organizational mediation of access and status in the field. The ideas proposed in the context of the AIC saw such an organization not only active in setting ethical standards for chemists and giving licenses for the practice as consultants but also in regulating the supply of chemically trained graduates (Eisenschiml 1922a), as an answer to the industry reconfiguring from wartime to postwar production.

Postwar Normalcy

Returning to peacetime industry, the cooperative networks between academia, industry, and the military and federal agencies dissolving or being put on permanent bases proved to be a topic once the war was over, during 1918 and the last event-tranche of 1921–1923. It can be considered a minor discourse since many currents of discourse were discussed in the context of others: the need to nurture the favorable atmosphere toward science created by the war (Campbell 1921; Williams 1921), legislation to protect the new industries that the war spawned, and the professional status of the chemist in a changed organizational landscape (Armstrong 1923). These discussions formed part of the larger discursive currents on Popularization and Legislation that were already part of the conversation prior to and during the war, yet they gained additional layers in the years after the armistice. Still, the return to Postwar Normalcy warrants a (small) topic of its own due to the way the role of the chemist and the place of research were sketched. The unemployment of chemists, having been let go from corporations or agencies that were scaling back from the "almost insane demand" (Bancroft 1922, 157), was frequently discussed. Discursive texts mirroring the efforts that ultimately culminated in the organization of the AIC show how this organization was conceived of as a professional body not only for the representation of economic interests but also to possibly regulate the supply of chemists available for industry and to counter the waves of lay-offs after 1918. Even though some only saw the postwar depression as only temporary (Bancroft 1922), others thought it imperative to restrict the supply of chemists, through educational campaigns (Eisenschiml 1922a; "American Institute of Chemistry" 1922). Still others reframed the problem by reconfiguring the chemist's subject position, calling for fields of activity for the chemist, who should not "get a job", but rather for them to go into businesses of their own ("Cogitations on the Chemist" 1922; Eisenschiml 1922c; "Shall I Educate My Boy to Be a Chemist?" 1922) or work as managers – thereby broadening the domain of the chemists as a professional group. In addition to translating the role of the chemist, the discussion on this topic concerned itself with the concept of research and its materialization in laboratories, warning of losing the ground gained during the war – not only in the perception of science and chemistry but also particularly in laboratories built and expenses set aside for research (Bancroft 1918; Parmelee 1921b).

Chemical Education

The only discourse appearing across all events was the discussion of chemical education, which touched on a series of questions: what are the goals of it, what should be covered, how to educate best, and should chemical education prepare for work in academia or industry? This discourse gives glimpses of the discursive construction of the concept of science, and the relationship between universities and corporations. One attribute of the education of all kinds of chemists, including industrial chemists and the chemical engineer, put forward repeatedly, was the breadth of education. Opinions on this were very homogenous as a broad consensus felt across all events. The education of chemists should be as broad as possible, with author after author underlining the necessity of learning the fundamentals of chemistry, instead of immediately specializing (Duisberg 1896; Richardson 1908; Coolidge 1921; Lovelace 1921; Bolling & Maze 1922). While there was a general consensus on the character of education that a chemist should receive, this issue serves to illustrate a wider point of contention, namely the relationship between universities and industries, the analysis of which helps us in understanding the discursive lines drawn between the university and the industrial corporation – a relationship that will play a role in the objectification of the laboratory, as well as the subject positions supplied for chemists by the discourse. In the battle for which kind of principles and applications were to be taught, some authors saw the deeper question of the function of the university, resisting the "seed of danger" held within the commercialization of the university (McMurrich 1907, 645). The necessity of education in fundamentals, then, was coupled with notions of the purity of university science and research (Remsen 1894, 534; McMurrich 1907, 645). Other authors denoted the ideal of the university as an institution of learning for learning's sake as being out-of-date, having made a new conception of the university as a utilitarian organization (Lovelace 1921, 357). This perspective understood chemical education in terms of supply and demand, with the university supplying the research workers that corporations increasingly demanded (Kellogg 1921), and the primary function of colleges and universities denoted as the education of chemists. After the war, concurrent with Popularization and the Return to Normalcy, the notion of "breadth" in education continued to be used, with some authors calling for an even broader education beyond the boundaries of chemistry so that chemists would be enabled to do more than laboratory work and thus find their place in marketing, sales, and management (cf. Bolling & Maze 1922). The executive's perception of the chemist as being employable only in a laboratory – chemists having been instilled at university with the idea that this would be their proper place – was seen as problematic for the chemist's engagement opportunities. Interestingly, for earlier events, the laboratory was constructed as the central place of education where the instruction of any kind of chemist should happen (Remsen 1894; Whitaker 1912a), thereby discursively consolidating the physical place where chemistry happens.

At the same time, the varying places of employment for chemists are reflected in the discourse. The industrial chemist – a label at first still seen as a "rather awkward designation" (Platt 1894, 37) – was discussed especially in relation to the realities of manufacturing work, the relationship between chemistry and engineering, and how well university education that was often seen as "pure" would translate to the shop floor (cf. Palmer 1908; Tingle 1908). The ability to do research, or "original investigation", was placed centrally in the array of abilities that chemical graduates should have, making them not only capable of but also valuable for industrial work (Tingle 1908). The different types of chemists, such as analytical, research, and industrial chemists (cf. Richardson 1908a) constructed in these texts further help as the first clues for substantiating the way the chemist was understood, where the lines between an analytical and a research chemist were drawn over time, and how they changed with the increasing importance of organized research in a corporate setting.

Concepts

To start the analysis of the discursive material with the analytical categories laid out above, let me briefly reiterate why concepts are important. "Concepts are culturally and historically situated frames for understanding social reality – ideas, categories and theories through which we understand the world" (Maguire & Hardy 2006, 13). The reality-parsing quality of concepts is what is most important to our undertaking here – when actors spoke of research, of the wonders of science, and of the domain of chemistry, they drew their knowledge of what science and research meant from shared, pre-configured, cultural notions, that is, concepts, instead of making it up each time one of these terms was used. Thus, in order to comprehend what science, chemistry, and research signified when discursive speakers evoked these terms, we have to analyze the ways they were constructed, what characteristics they were attributed with, and how they changed over time. The choice to analyze the concepts of science and (industrial) research on the one hand, and chemistry on the other, was both theoretically and empirically informed. For one, ideas about what science is, what doing research entails, and what products chemistry could conjure have necessarily had to play a part in the discourse surrounding the early laboratories. Yet the analysis had to be shaped openly and attentively enough in case other concepts came to the fore, thereby influencing the discursive trajectory, which proved not to be the case in the empirical material. In the following, I will first present how the interlinked concepts of science and research – research as doing science – were constructed, and how industrial research emerged discursively around 1912. Then, chemistry itself will be scrutinized regarding its meaning, applications, and efficacy.

Concept 1: Science, Research, and Industrial Research

Since industrial research laboratories were places of research – potentially even scientific research – it is imperative to know the prevailing conception of science and research at the time of the laboratories' inception. Early on, science was established to signify a quest for truth (Goode 1895; Carhart 1895, 397), but not just any kind of truth – the truth of nature. It was repeatedly stressed how the scientist's work meant discovering nature's facts and the underlying laws (Newcomb 1895; Bauer 1909), which was often seen as a struggle with nature to uncover its mysteries: "... literally numberless are the conditions under which matter is ever anew compelled to yield up its secrets" (Emich 1900, 444). The process of uncovering – the struggle for discovery – meant getting in touch with nature and as closely as possible. Remsen denounced the methods of philosophers to "stand aloof and speculate", whereas (natural) scientists learned that to uncover the truth they would have to come in close contact with nature (Remsen 1894, 533; Coulter 1900, 281), and that meant touching, observing, probing, weighing, and measuring (cf. Bauer 1909, 188) the elements, be they physical, chemical, or otherwise. But how would science go about wrestling answers from nature, once it had gotten into close contact? The practice of science was discursively constructed as finding facts through experiments, observation, and the gathering of evidence. Science consisted of "weighing evidence and stamping each statement with an index of its reliability" (Scripture 1895, 350). In fact, the evidence-based nature of scientific practice can be seen as its core tenet (cf. Brooks 1895; Bessey 1896).

The close relation of science to nature and the scientific practice as gathering evidence through observation and experiment came to be codified in the concept of the scientific method:

> The scientific method is therefore that intellectual process by which facts are recognized, accumulated and arranged, hypotheses framed, tested and exploited and conclusions drawn, verified, accepted and applied where they may seem best to fulfil their function in the enginery of social progress.
>
> (MacMillan 1895, 538)

Science soon came to be identified with the scientific method. When discursive speakers wrote about science, they generally referred to the ensemble of practices subsumed in the method. In fact, the scientific method came to be celebrated to the extreme, as the "most satisfactory and satisfying things in the possession of the human mind" (Richardson 1911, 514), the most marked change in educational history (Carhart 1895, 394), and the means to eliminate errors (Turner 1900, 358). Soon, proclamations were made that no other method existed for finding truth, for making nature reveal its ways: "Revelation and prophecy are thus fruits of science. It may perhaps be said, with truth and literally, that we to-day know no other method of either revelation or prophecy" (Thurston 1902, 402). Indeed, the use of the scientific method can be understood as not only finding laws underlying facts but also rendering supremacy over nature possible (Halstead 1895a; Little 1909; Crowell 1909), which further changed the relationship of both, with science now assuming a dominant position over the natural world, the limits of which were unknown: "Science knows no ultimate limits beyond which she may not go … nor has nature yet thrown up her hands as a signal that she no longer resists the uncovering of all her treasure" (Carhart 1895, 402). It is not surprising to see how these ascriptions of the scientific method changed with other discourses coming to the fore. Around 1912, during the height of the efficiency craze, the scientific method was referred to as being more efficient than the older ways (in industry), whereas during WWI, German military may be acknowledged as being a product of widespread application of the scientific method, though in an unethical way. While texts outlining – or celebrating – the meaning and practice of science were more frequent for early events, the basic relationships between truth, nature, and science described here did not change: science meant using the scientific method to uncover nature's truths.

From Science to Research

Now that we know the contours of what science signified for the seven events, we need to know how research was connected to science. It can be argued that research soon became synonymous with science (Doolittle 1902; Crowell 1909),[2] potentially with more emphasis on the practice of it – research thus meaning *doing* science, as "diligent inquiry or a search after a desired object" (Weens 1917, 478). Thus, research was cast in the same relationship to truth and nature as science was: the method for finding truth (Walker 1911, 17), and the conquest over nature (Robertson 1917, 375). A key element in the construction of research as the practice of science was the notion of original investigation (Remsen 1894; Carhart 1895; Baker 1900; McKeen Cattell 1902). The originality involved in doing research thus served as an important juncture for establishing the characteristics of the scientist, as well as stressing the creative forces involved in the practice of science as the "fountainhead of new knowledge, the vital stimulus of industrial growth, the originator of new industries and sustainer of old" (Swan 1901, 673). In a parallel to the construction of science, the concept of research connected to other discourses in various ways over time. It was often discussed with reference to the research landscape, especially in the context of the emerging dual function of the university in teaching and research, with teaching duties often strongly impinging on the time available for original work. Furthermore, as outlined, adequate financial and institutional support for research was repeatedly questioned, with many ideas and models put forward (Swan 1901; Clarke 1902). Research, pitched as the solution to conservation problems (Sadtler 1909), or later preparedness (Whitney 1916; Bogert 1917b), with lowering profit margins and the quest for efficiency, gained an additional dimension that exceeded simply being science put to action. Speakers increasingly constructed research in opposition to the traditional ways of doing things, the rule-of-thumb methods, the old haphazard days of trial and error (Little 1913, 644). The organized character of research was regularly emphasized, as well as the slow, methodical, and planned character of doing science (Talbot 1912; Roeber 1913; Clarke 1916) and the idea of research as a cooperative activity of scientists in a group (Roeber 1913; Bacon 1917a). The concept of research as doing science to subjugate nature serves as an important element of the construction of the industrial laboratory, since – and this is one of the key results of this analysis – both science and especially research were always located within the confines of a laboratory. This illustrates the discursive tendency of locating concepts, with the place chosen and changes to such locations giving important clues as to the discursive ruptures and changes in the organizational field.

With regard to science and research as discursive concepts, two more important dimensions need to be sketched to understand their materialization: the reasons for doing science, and what exactly came to be signified when *industrial* research was talked about. What were the advantages of science, the claims made for research – why do it at all? Science and research were constantly put into the context of societal progress and national welfare, and these were generally connoted as material progress and welfare. Depictions varied from the grandiose – science bringing about "glorious change" (Halstead 1895a, 203) or an "uplift of humanity, an advance in civilization which cannot be described or measured in words" (Walker 1911, 17) – to more concrete claims: "In order not to become a mere code of observed and classified facts, a thing to interest men by its intrinsic perfection but of no other earthly use, a science must find practical applications which increase the comfort, pleasure or wealth of mankind" (Richards 1903, 66).

Comfort, pleasure, and wealth were claimed to be reachable through science not only on a national level (Carhart 1895; Thurston 1900; "Research Institutions" 1913; Roeber 1916) but also for each citizen – those hungry fed by science in agriculture, those hurting relieved thanks to chemistry and biology, those freezing relieved by research into fuels. It was made clear how science could impact daily lives in a concrete and tangible way. Yet it was the applications of the fruits of science and research that brought these improvements – on the one hand, the application of scientific facts to industrial production in existing industries, and on the other science laying the foundation for new industries (McMurtrie 1901, 441; Thurston 1901; Little 1907; Crowell 1909; Bushee 1911; Whitaker 1912a). Over time, this was even described as a heavy debt that industry owed to science (Carty 1916). In this way, the discoveries made by science in the practice of research were strongly ascribed not only to abstract ideas about the progress of civilization and higher human ideals but also to concrete achievements that could be measured and felt. Yet these discoveries, facts, and laws would not bring about change by themselves, instead, a clear path was laid out: it was through applications in industry that these fruits could be reaped (cf. Tucker 1908; "Science and Industry" 1917). Interestingly, when the work of great scientists was referenced, it was usually done in relation to the applications of their work, meaning the discoveries they brought into specific industries. Since the discursive corpus analyzed here covered chemistry in all its varieties, the scientists mentioned were primarily chemists. Liebig was lauded for his advancement of agricultural methods (McMurtrie 1895; W. Walker 1911), Pasteur was celebrated for his discoveries in bacteriology and the distinctly application-oriented focus that his research work took (Halstead 1895b; Maclaurin et al. 1916; Jacobson 1916; Ogden 1918), while the work of the forefathers of electrochemistry – most mentioned were Faraday, Maxwell, and Kelvin – was put in direct relation to any advances in the harnessing of electric power for industrial use (Pupin 1895; Bauer 1909; Maclaurin 1916). Furthermore, their triumphs – or the triumphs ascribed to a depersonalized concept of science or research – were constantly asserted as being so well-known that they hardly needed elaboration, and this proved to be true across all events, whereas one might have expected the prevalence of such knowledge to increase over the period analyzed (Carhart 1895; "Address of the President" 1900; Wiley 1902; Montgomery 1908). Thus, science and research were clearly constructed as the way to material progress and welfare, which could be attained by industrial applications, with the service of science already acknowledged in 1895 as being so great and widespread that it hardly needed to be elaborated upon.

From Research to Industrial Research

But what about *industrial* research? How was this descriptor connoted: merely as a different type of research, or as something completely different? As it turns out, actual invocation of the term "industrial research" only occurred with the fifth event, 1912. Before, circumlocutions existed that signified scientific research as not happening within the confines of a university, mostly in relation to the actors who did the research. In this way, Richards (1903) speaks of investigators in corporate plants, while by 1908 several authors wrote about the "industrial chemist" – a chemist employed by a corporation (Teeple 1908; Tingle 1908). These chemists were often named in discussions on the proper education of chemists, and how their knowledge and skills for original investigation would transfer into a corporate environment. By 1912, "industrial research" was actually called that, and the discussion broadened. Authors started to acknowledge that research happened in industry and was indeed necessary for industrial progress (Cottrell 1912; Whitaker 1912a) as the "guiding light of chemical industry" ("Industrial Research" 1911). Furthermore, the establishment of corporate laboratories and university departments for industrial research were regularly announced (e.g. "Facilities for Industrial Research" 1911; Benner 1912), indicating a certain normalization of the practice of industrial research located in special corporate or university laboratories. Similar to "regular" research, a foundational role was ascribed to industrial research, in the development of manufacturing and engineering industries ("Facilities for Industrial Research" 1911, 797): finding the fundamental truths or basic principles that need to be understood to progress (Little 1913, 644). Regarding the relationship between industrial research and chemistry, it was repeatedly linked to applied chemistry (Walker 1911; Little 1913). Similar to Germany being the most advanced in the science of chemistry, the German chemical industries were frequently named when the triumphs and advantages of industrial research needed to be recounted (Roeber 1913). For the next event, 1917, many of the attributes attached to industrial research were repeated or deepened. The announcements of research laboratories and departments continued (e.g. "Scientific Research and the Electrical World" 1917), while the discussion gained a theme related to Preparedness and the war effort: the question of how to best organize an industrial laboratory (on a corporate, as well as a national level) (Mees 1916; Norton 1917; Hill 1918). The achievements of industrial research were communicated and celebrated (Little 1916; Bacon 1917a), especially by speakers hailing from the pioneer laboratories (Carty 1916; Reese 1918), as its relation to national welfare became an increasingly important point even with the end of the war in sight (Hill 1918; Root 1918), since German supremacy through research needed to be countered by the development of America's own facilities. For this event as well, industrial research was described in close relation to the applications of science and applied chemistry, or in opposition to its pure form (Carty 1916; Sharp 1917) – descriptions that could also be found in the set for texts of the last event, 1922. The industrial research laboratory continued to be a theme of intense discussion (Parmelee 1922; Wardenburg 1922; Little 1923) located in corporations and working on commercial goals (Hyde 1922; Armstrong 1923).

As can be seen, the term "industrial research" as a name for the activity was only fixed when the actual practice of establishing laboratories in corporations took off. But instead of being described as something completely different, industrial research was clearly posited within the same network of associations as research, the practice of doing science with similar – sometimes grandiose – claims to utility and necessity, as well as similar attributions in the way it was conducted – by groups of trained scientists, in a laboratory setting.

What remains, then, is to take a look at the notions of pure and applied chemistry that were increasingly invoked in discussions of industrial research. The historically elusive distinction between both types was indeed drawn and frequently discussed, and this happened over the entire period of analysis. While the signifiers changed – some authors proposed alternative categories to pure/applied, such as "time-consuming"/"resources-consuming" or science/technology (Ogden 1918) – one generally recurring detail was the assertion that both types only differed in motive, and in no way differed in method. That is, both were exemplars of the scientific method, while the goal – knowledge or profits – might vary (Walker 1911; Clarke 1916; Carty 1916; Parmelee 1921c; Hyde 1922). Of course, the precise construction and change of meanings of pure and applied science (or chemistry) is not of central interest here. Instead, what soon became apparent and serves as a further illustration of the tendency of locating in this discursive fragment, is that distinct, separate "places" were assigned to pure science and its applications, a distinction already visible in the placement of industrial research. The university was seen as the locus of pure science – "the natural home of pure science and of pure scientific research is to be found in the university, from which it can not pass" (Carty 1916, 515) – while applications could happen at universities, but were largely done elsewhere: in industry, in consultancies, in government laboratories, at technical schools, etc. Locating science/research according to type grew more distinct and clear over time (cf. the differences between Bogert 1913; Wilbur 1916; Kellogg 1921). It is interesting to note how the differences between both types were not constructed according to who did the research and what their scientific credentials were, or what eventually happened with the knowledge gained in the process. Instead, profit, or more concretely improved products and processes – being the overarching motive ascribed to industrial organizations – was the reason research located in industry, which came to be known as industrial research, was aligned with applied science. On the other hand, the advancement of knowledge was the university's goal, and in this way research done there would stay "pure". Even with this tendency of "placing" a concept, the split of research and industrial research, as well as the opposition of pure and applied research, was not absolute. This is illustrated by departments of industrial research opening at universities and the continuing discussion of research funding and industrial education, as well as claims of both types of research being done in industry (Flinn 1921), and calls for the need to understand "fundamentals" which could only be solved by doing pure science work (Wilson 1922). "Fundamentals" nicely illustrate the temporalization of pure science by the later events, with the distinction of pure and applied research gaining a recognizably temporal dimension in the linear relationship postulated: pure comes first, and applied after. The progress of applied science was seen as heavily dependent upon advances made in the pure sciences (Mees 1916; Jacobson 1916; Stieglitz 1917a; Raiford 1917; Withrow 1917; Coolidge 1921; Bancroft 1922). Pure research was now portrayed as slow, not to be planned for, and a gamble (Hyde 1922; Derick 1922), further strengthening its organizational location at universities, which would supply support for goal-less research for a longer time.

Thus, locating a concept in time and space happened not only on the level of science/research, but also in regard to its types: research happens in a distinctly physical place, a laboratory, but what kind of research is done depends not on the output or research worker, but rather on where that laboratory is located. University and corporation served as providers of the overriding motive, truth/knowledge or applications/profit, marking the difference between scientific and industrial research. Yet regardless of the organizational locus, research meant practicing the scientific method in a pioneering effort for progress, wealth, welfare, and nation.

Concept 2: Chemistry

Understanding the importance of how the concept of chemistry was constructed in the discursive reality surrounding the early industrial research laboratories, one need only remember that the pioneering laboratories were chemical laboratories for the most part, and that chemists were the largest group of scientists in employ in corporate laboratories. During the analysis, it soon became clear that the attributes credited to science as a whole were also invoked when talking about chemistry. Chemistry was thus the road to truth (McMurtrie 1899), welfare (Richards 1903), and wealth (Clarke 1902, 1916; Little 1908; Whitney 1921; Burgess 1923), permeating "every field and phase of modern life ... the public mind has hardly yet awakened to its full significance" (Richardson 1908b, 810). Furthermore, chemistry was not only constructed as a science among others but also as the fundamental science without which others – such as physics, pharmacy, or engineering – could not have made advances (Wiley 1902; Toch 1909; Bancroft 1922). Chemistry's prime position lies in its object of study – matter: "... *chemistry is the science of the transformation of matter*. ... we find at every turn in life that chemistry is in demand to aid man in his effort to assure to himself a safe, scientific control in the supplying of his own needs, where nature ... has shown the same impersonal indifference as to his wants, his survival or destruction ...!" (Stieglitz 1917a, 2096, emphasis in original).

The idea of not only studying different forms of matter but also transforming matter – working closely with atoms as the basic building blocks of nature itself – occurs frequently (Vaughan 1912; Scholes 1918; Bancroft 1922), with the productive aspect of chemistry thus being stressed as not only an analytical (breaking down) but also a synthetic (building up) science (Brogdon 1912). Similar to science and research, in the discourse claims regarding superiority over nature were made, superiority that chemistry could grant precisely because the properties of matter were its concern. While some referred to the reproduction of nature in a laboratory setting (Little 1909), others went so far as to claim that chemistry could surpass nature and make natural processes less "clumsy" (Matthews 1902, 429). Yet the chemical focus on tiny particles of matter also carried with it a difference that will be further explored when talking about the chemical engineer. This could be summed up as being a size difference: the chemical processes, reactions, and experiments happening in a laboratory, working with minute quantities at a time, would need to be scaled up once translated to the shop floor, and the chemical engineer was precisely the natural actor required to undertake this (Whitaker 1912a; "Where are the Leaders?" 1918).

The Utility of Chemistry

What were the reasons for doing chemical research? Again, the many practical uses of chemistry were praised, especially in its role as the basis of the industrial production of wealth, and not limited to the chemical industries but others as well, such as agriculture or manufacturing in general (Mason 1895; Clarke 1902; Richardson 1908a; Duisberg 1912; Raiford 1917). The part chemistry played in fueling industrial progress was deemed so well-known that it hardly needed to be mentioned (Wiley 1902, 846).

The fundamental nature of chemistry, dealing with matter, came to be referred to in claims about the use of chemistry in the industries, underscoring why the applications of chemistry, and with it, the employment of the chemist, should know no bounds. Because all industry dealt with the matter in some form, and chemistry was constructed as the expert science on all kinds and transformations of matter, it followed that all industry was chemical in some way: "It is next to impossible to imagine the existence of an industry in which chemical reactions or considerations, either directly or indirectly, do not enter" (Thompson 1913, 800). Moreover, authors claimed that every technical problem had a chemical part to it (Tingle 1908), and that chemistry and mechanics should be regarded as the "first parents of technical industry" (Louis 1917, 742). Unsurprisingly, chemistry's utility was constructed and translated in relation to prevalent discourses. During the height of the Conservation Movement and the efficiency craze, chemistry (and chemical research) became the means for improved accuracy – scientific exactness, that is – and the countermeasure against the prevailing trust in empiricism and the rule-of-thumb on the machine shop floor (Richardson 1908b; Little 1911; Talbot 1912; Clarke 1916). During the preparation for WWI, the wonders of chemistry were praised for this quest, especially citing the German system of support for science and industry. The German system of science and chemical industry were models exalted across the whole period, as the uses of chemical research gained in argumentative power through frequent reference to Germany.

The reasoning employed was quite simple: Due to a great appreciation for science in German society and thus industry, the fruits of chemical research were used to the greatest extent, which made the German chemical industry the world leader. These claims were spread not only through chemistry PhDs returning from study in Germany but also through addresses by eminent German figures (Duisberg 1896) and reports on the German way of doing things (e.g. "Chemical Industry in Germany in 1901" 1902). The German head-start in chemical manufacturing was widely acknowledged, as well as the fact that the American industry could not compete yet (Clarke 1902; Comey 1912), mainly because the United States lacked an equal appreciation for science, chemistry, and the chemist. Germany not only served as a paragon of what a collective devotion to science could yield (cf. Walker 1911, 26ff; Thompson 1917a) but also the fact that research chemists were employed in laboratories within corporate confines in the German chemical industry was seen as proof that this organizational entity would bring direct benefits to a chemical corporation:

> Valuable information is on record showing how, in numerous cases, the research laboratory has been a tremendous profit to industry. In some cases the research laboratory is devoted almost entirely to the development of new processes and products, and it would appear that the Germans have most successfully applied this method, and that their commercial high standing in chemical manufacture has been more due to this than to any superiority in methods or economies in manufacturing.
>
> (Thompson 1913, 804)

The magic of industrial research, practiced by the research chemist, was thus constructed as the key to German dominance (Whitney 1916; Norton 1917). Of course, having become the enemy in a war that revealed the horrors of modern science, Germany's scientific, industrial, and military achievements could not be celebrated or used as a prototype to aspire to anymore. In some discursive texts, Germany was still constructed as a nation highly valuing science and industrial research, yet at the expense of morals: "the object was selfish and nothing more nor less than was expressed by the motto 'Deutschland über alles'" (Nichols 1918, 771), with the men of science having fallen under the spell of the debased character of the German leaders (Ames 1918, 402; cf. also Washburn 1916). While this discursive figure remains fragmentary, it may serve as an explanation of how the German model of scientific education and industrial research continued to be espoused at a time when Germany was damned as the aggressor and scapegoat of WWI.

To summarize, chemistry and chemical research were constructed in a similar way as science and research, with chemistry claiming a foundational role in all other scientific development due to its proximity to nature, and the study of matter also being the reason why supremacy over nature was a claim chemists made from their laboratories. Furthermore, the utility of chemical research was found in its myriad applications to various branches of industry, welfare, and wealth, further underscored by Germany's successes in industrial chemistry since the second half of the nineteenth century.

The last question remains: Whether or not similar oppositions of pure and applied chemistry can be found. First of all, while a distinction between pure and applied chemistry is indeed constructed, a plethora of other terms for the varieties of chemistry are used as well, from regularly occurring terms such as "industrial chemistry" (Gilfin 1900; McMurtrie 1901; Messel 1912; Stewart 1912) and "industrial chemical research" ("Industrial Research" 1911), to less frequently used descriptors such as "technical chemical research" (Comey 1912). An interesting division opened up around the fourth event, 1908 onward, when chemistry was increasingly distinguished as either analytical chemistry or research chemistry. Sometimes aligned with industrial chemistry (McMurtrie 1909) or mere control work in industrial plants (Richardson 1908b), analytical chemistry was also seen as a servant of other branches or a less developed type of chemistry (Talbot 1912) and put in direct opposition to chemical research (Ferguson 1912; Thompson 1917a). This evolving distinction may serve as a hint of the increased practice of doing research in industry, instead of using chemical knowledge for testing and control purposes, and will be further discussed since both types of chemistry were most often put in the context of actors – that is chemical researchers and analysts – instead of remaining in the abstract realm of concepts. The term "applied chemistry" only occurs in the later events, and in 1908 only in relation to the dangers of specialization that would make chemical practitioners lose touch with each other. It was only later on that opposition was constructed, and it followed similar lines to the one of pure and applied science, with applied chemistry happening in manufacturing on the shop floor – in a corporate setting (Wesson 1912; Withrow 1917). Several authors stressed the similarities between pure and applied chemistry as being merely phases of the same science, the only difference existing in their purposes (Walker 1911, 27).

The goal of chemical research was subject to localization, as happened with pure and applied science – corporations supplying an overarching purpose of profit, products, and practicability, while chemical research at universities, its pure variant, sought for basic principles and an understanding of nature and truth. It must be noted, though, that pure and applied science were far larger topoi within the discussion of types and places of science, with chemistry in turn yielding descriptors and oppositions that went beyond the pure/applied dichotomy, such as the emergence of analytical and research chemistry sketched earlier. What can be asserted, though, is that applications of chemistry – as applied chemistry – to industrial purposes were highly visible and constructed as the reason for industrial success.

Now that the discursive meanings of science, (industrial) research, and the varieties of chemistry were analytically reconstructed, interpreted and reflected upon within the contexts of the organizational field and the historical genesis of science, it is important to learn in what way they were materialized into objects.

Objects

As Munir and Phillips write, objects are "part of the practical realm made sensible by discourse. ... Put another way, when a concept is used to make some aspect of material reality meaningful, an object has been constituted" (Munir & Phillips 2005, 1668), highlighting the essential connection that exists between concepts as categories guiding thought and perception, and the way they are materialized in concrete, physical objects. Objects are made meaningful by discourse, and the meanings associated with these parts of the material world are subject to frequent changes. For our purposes, then, it is important to understand in what ways the concepts already analyzed were materialized, what parts or specific characteristics were left out, and how attributes of an object – or the whole object itself – changed or had to change to fit the discourse. Whereas a variety of historically situated concepts emerged in the preceding analytical category, only one object proved central within both the discursive material and the theoretical avenue of inquiry, an object basal not only in the realm of science but also as an organizational entity; yet an object that was still subject to the discursive effects of place and time.

Objects: The Laboratory

The laboratory emerged as the central element of the analysis. This may be of no surprise to some since the main theme of this book surrounds the industrial research *laboratory*, yet most of the work in the history of science surveyed before is concerned with the first part, industrial research, and trying to make sense of how research made its way into industry – thereby taking its concrete organizational form as a laboratory for granted. In this way, expectations at the start of the analysis veered strongly toward understanding science, research, the industrial chemists' plight, and so on. But during the analysis, the importance of the laboratory as an object and place of science became more and more notable, findings that are further underscored by insights from historical studies on the genesis of the laboratory – a place that we today expect to be the logical, even natural locale of chemistry, but a place that was hardly as fixed and taken for granted during the outgoing nineteenth century. Thus, in the following, I will describe the discursive steps that made the laboratory an object of the materialization of science and the place of research – constructed not as an element of universities, but one that was transferable into an industrial context. Moreover, with the value of research and applications of science in industry being well-known, the reasons for establishing research laboratories were seen within this applicability, becoming a place where experts – chemists – were under employ. Finally, changes and trends in object construction will be outlined in the emergence of ideas on laboratory types, laboratory scale as related to the factory, and the way that the pioneering laboratories were discussed during the later years.

The Laboratory as Birthplace of Facts and Place of the Scientific Method

Why can the laboratory be seen as a materialization of the concepts of science, research, and chemistry? The attributes of science carved out above – the scientific method, progress and welfare, the subjugation of nature, and applicability in manufacturing – came to be explicitly associated with the laboratory. Experimentation as the core element of the scientific method was seen as only possible within the confines of a laboratory since it offered a stable environment and the equipment necessary to unlock nature's secrets:

> You will have guessed already the difficulty with the chemistry of the Greeks. It amounted to nothing more than speculation. The hypotheses were never tested in the workshop of science – the laboratory; they remained at the last what they were in the beginning – unproven products of the imagination.
>
> (Richardson 1908b, 804)

It was this possibility of close but controlled contact with nature (Remsen 1894; Coulter 1900), the subject matter of the natural sciences, that made the laboratory the physical manifestation of the scientific method (Carhart 1895, 394f). Understanding the scientific method as the way to establish facts about nature in the search for its underlying principles meant that laboratories thus became the "birthplace" of new facts (Richards 1903, 60), and this was asserted not only for facts about nature but also in the search for facts that could solve problems in industry ("Facilities for Industrial Research" 1911, 798). Many authors lauded the value of laboratories for purposes of research (Howe 1902; Thurston 1902; Little 1909; Mees 1917a), in this way further strengthening the discursive association of the practice of science with the laboratory. Most of the texts that built these associations came from the early events (1895, 1900, 1902, 1908), whereas later events eschewed the need to outline the purposes of laboratories as places of science. Such a reduction in argumentative frequency can be interpreted as the materialization of the concept of science in the laboratory acquiring a certain acceptance among speakers, as the characteristics of fact-creation and experimentation in the laboratory were now so strongly associated with the object that they need not be reiterated any further, with the discussion now shifting toward laboratory practices, equipment, costs, and other themes. Interestingly, the laboratory was not placed explicitly at universities – that is, no clear link was constructed that would mean this object would need to be housed in the organizational surroundings of the academic world, a notion that becomes clearer when looking into the relationship between teaching and research surrounding the laboratory. From the earliest event onward, the laboratory was connoted as a place of teaching, with its essentialness to learning the methods of science recounted time and time again (Remsen 1894; Carhart 1895; Clarke 1902; Whitaker 1912a), with some authors claiming that laboratory instruction had revolutionized science learning (Coulter 1900). But next to its role in the teaching of scientists the laboratory was always also a place of investigation, its dual function so clearly noted (Remsen 1894; Baker 1900; Clarke 1902; Whitaker 1912a; Sedgwick 1913) that its value to research was claimed as self-evident: "… the use of these laboratories, not for purposes of investigation, for which their value is unquestioned …" (Howe 1902, 762). This is further illustrated in early discussions of the research landscape, where calls for research endowments were made – these calls clearly meaning the endowment of laboratories:

> On several former occasions I have advocated, as the most urgent need of science, the regular endowment of research. … I look rather to the establishment of institutions, wherein bodies of trained men should take up, systematically and thoroughly, the problems which are too large for individuals to handle. … in the form of a well-built, well-equipped, and well-endowed laboratory ….
>
> (Clarke 1902, 136)

While the laboratory in its teaching function was clearly constructed as a part of the university – the natural and unquestioned place of the education of the scientist – the recurring emphasis on the dual function of teaching and investigation, often elaborated upon in the same breath, made it possible for the laboratory to not be seen as distinctly academic, an entity of the university, but simply as the place where research happens, a place that could probably be moved to other organizational boundaries. The early event-tranches constructed the laboratory as the physical manifestation of science and the place where new facts are found, while underlining its role in teaching and research. Within the larger universe of discourse reconstructed above and in materializing science in the form of the laboratory – and not the university, the mind of the scientist, nor any other way – the grand promises of science for progress, welfare, and happiness can be considered extended to the laboratory. But in this extension the construction of the object was not complete – to further understand how science became materialized in the confines of the laboratory, we need to consider the reasons for establishing research laboratories and the people who worked inside.

The Why and Who of Laboratories

Understanding the claims made in the reasons for building up laboratory facilities in industrial corporations is important, first and foremost for examining whether reasons for doing research were mirrored here. As it turns out, texts arguing for the need to establish (and further endow) laboratories did so mainly from a vantage point of the applicability of the facts generated within, yet few overarching, recurrent argumentative figures emerged. Discursive figures included reasons for building up laboratories that were sometimes described in the rather abstract terms of progress and prosperity ("The Status of Electrochemical Industries" 1908; Sharp 1917), or as the "experimental and exploring arm of the up-to-date industry" (Crowell 1909, 563; cf. also Whitney 1911; Thompson 1913), pioneering imagery being a recurring motive in descriptions of science and the scientist. The applicability of the laboratory's findings to specifically industrial endeavors was central in many of the texts analyzed (Richards 1903; McMurtrie 1907; Whitaker 1912a). Others focused on economic benefits in terms of profits or the extension of the boundaries of business (Thompson 1913; Bacon 1917a), or posited that a laboratory's benefits were not measurable in monetary terms (Whitney 1911; Sharp 1917). Unsurprisingly, discursive shifts played a part in the way that the laboratory as object was reasoned for. Specialization assumed only a minor role – as Thurston (1902, 415ff) argued, due to the differentiation into many sciences, it followed that many specialized laboratories were needed. Between 1908 and 1912, the laboratory became associated with the search for increased efficiency, whereas Willis Whitney of GE even claimed that the laboratory was an "extension of the principle of maximum efficiency" (Whitney 1911, 429).[3] Similarly, when the war came, the key to a successful military campaign – as a problem of applied science – evidently lay in laboratories (Bogert 1917b, 1010), whereas the German "dreadful efficiency" could be traced back to chemical research laboratories (Noyes 1917). Overall, even though no single, dominating discursive figure giving clearly outlined reasons for establishing a laboratory could be found, the general tenor is similar to the ascriptions made to science and research, yet often in less abstract or grand terms, and not by invoking societal progress or national wealth, but rather reasons such as staying ahead of the game (Thompson 1913), profits, or a better understanding of manufacturing basics, as mediated by the relevant discourse of the day.

A key point in the historical research arguing that the industrial research laboratory represents a fundamental shift in corporate strategies for innovation refers to the employment of scientists in these laboratories, with manufacturing organizations moving away from purchasing products and processes from inventors. While the specific subject positions constructed in this discourse will be discussed below, the question of who would – or should – work in a laboratory remains an important part in the materialization of the laboratory as an object. In analyzing this relationship, two distinct argumentative periods could be distinguished: an early one, ranging from roughly 1902 to 1912, and a later group of arguments emerging with the end of WWI that culminated around the last event, 1922. First, during the early events, the laboratory was clearly constructed as the place of employment of chemists. This happened through the description of the discoveries and triumphs made by chemists in laboratories (Chandler 1900; Matthews 1902) or by depicting laboratory work as a "tool of the trade" of chemists (Withrow 1911, 626), with the (company) laboratory seen as a typical first place of employment for chemists after graduating from university or college (Richardson 1908a; Burgess 1911). Furthermore, distinguished chemists and the forefathers of chemistry were often discussed in relation to their laboratories. Berzelius, for example, succeeded despite lacking proper apparatus: "Berzelius, one of the greatest chemists, pursued his simple researches in a laboratory established in a kitchen …" (Wiley 1902, 842), whereas Liebig's laboratory at Gießen "stands for a new epoch in scientific investigation and instruction" ("Justus von Liebig and the First Laboratory" 1909, 619). Especially after the first decade of the twentieth century, chemists in laboratories were discussed more and more in relation to manufacturing, the shop floor, and industrial production. By then they came to be seen as a necessity – "A laboratory is just as essential to a factory as is an office, and the chemist is just as necessary as the auditor" (Skinner 1911, 308) – and the way that traditional, empirical methods could be replaced with new, chemical knowledge (H. Skinner 1911; Brogdon 1912). Even though the fundamental element in the materialization of the laboratory was its relation to the scientific method, in the texts surveyed here the laboratory was constructed as a place for chemists and not scientists as a broader category, with the chemist obviously being a practitioner of science. This is due to the selection of texts that are primarily discussing chemical (and related) laboratories, and only a few speaking of scientific laboratories per se – hence the chemist and not the more generic scientist being placed in the laboratory. Thus, in the first period a strong link was established between the laboratory and university-trained chemists working in groups in these laboratories. This link was still present when discursive references shifted to Postwar Normalcy (Garvan 1921; Layton 1921), but the discussions on unemployment, the role of the chemist, and the postwar reconfiguration of research institutions found their way into how the laboratory was materialized. Many industrial speakers problematized this strong link between laboratory as the natural place of the chemist, urging chemists to broaden their focus from research and analysis to administrative and executive work. The "directional emphasis" (Redman 1922, 292) in chemical education toward laboratory work was criticized in efforts to further popularize and strengthen the role of the chemist in the sphere of learned professions (Bolling & Maze 1922). Whereas earlier on the laboratory was the way for chemists to enter into industry, it was now seen as confining and a reason for rising unemployment (MacDowell 1922; O'Brien 1922). The second period's emphasis on getting chemists out of the laboratory further underscores one of its central attributes, namely being staffed by groups of chemists as experts and practitioners of science. Indeed, the only time an inventor's laboratory (as opposed to a chemist's or a scientist's) was mentioned, it happened with the caveat that while such work may be revolutionary, transforming it into commercial results would only work "following strictly the rules of logics" ("The Status of Electrochemical Industries 1908, 2), possibly hinting at the methodical and foundation-revealing aspects of science as opposed to an inventor's work process. The laboratory, then, was strictly in the hand of chemists.

Scale and Scope: Laboratory Relations and Types

The interlocking elements in the laboratory becoming an object as a specific materialization are now laid out: The laboratory serves as the physical manifestation of the scientific method, wherein close contact with nature generates new facts. The dual function of teaching and research serves to dissociate the laboratory from the university, opening it to possible translation toward corporations – especially since the laboratory, staffed by expert scientists, not only produces facts but also can aid in applying these to industrial production, generating more or less tangible benefits in profits, the extension of markets, and staying ahead of the competition. These attributes remained fixed throughout the seven event-tranches analyzed, with only minor shifts in discursive references, for example, laboratories for efficiency or preparedness. Yet several changes occurred over the years. The first concerns the external placement of the laboratory, in a display of the discourse specifically locating concepts and objects. External, since this time it is not science that is placed within the walls of a laboratory, but the laboratory itself that is placed in relation to other elements of the industrial organization. The changing relationship between laboratory and factory could best be characterized by the notion of scale. Starting with the fourth event, 1908, authors began discussing the challenges of transporting the findings of laboratory research to the shop floor: "We often hear of the success of a laboratory method, and its failure when applied on a manufacturing scale" (Parker 1909, 2).[4] Such failures were initially attributed to a lack of laboratory equipment and acquaintance with "the raw materials of the industry" (Sadtler 1908, 37). By later events, solutions were proposed to replicate industrial-scale reactions and processes (Bogert 1913; Little 1913; Sharp 1917), partitioning the perceived "distance" between laboratory and factory into separate stages, from laboratory to test-tube to small-scale plant and to regular manufacturing scale (Bacon 1917b; Baekeland 1917; Sharp 1917; Hill 1918; Teeple 1922) a frequent suggestion. Three specific parts in these discussions are of special interest here regarding the object "laboratory". First, the relation of scale constructed between laboratory and factory. As it turns out, according to discursive speakers laboratories did not fail because of the shortcomings of the scientific method, but rather because what happened in a laboratory was so small – atoms, molecules, mere grams of certain substances, instead of the tons involved in industrial production: "It makes an enormous difference whether you are manufacturing by the ounce or by the ton" (Baekeland 1917, 1021). This applied not only to substances but also to apparatuses and procedures:

> It must be remembered that while many discoveries are made by laboratory experiments on a small scale, many of these are not applicable to industrial exploitation … . A chemist in his laboratory may effectually close a flask with a cork, a similar procedure does not apply at all to a spouting oil well.
>
> (Sharp 1917, 168)

As can be seen, the discussion of the laboratory's scale in relation to the factory that intensified from 1912 onward did not deny the laboratory's credentials as the materialization of science and supremacy over nature, but merely repositioned it in the industrial organization. The second part of interest in this discussion concerns the subject positions involved, namely the industrial chemists and chemical engineer. Laboratory scale began to be discussed with debates on the organizational home of chemical engineers ceding and the founding of AIChE (1908), a string of arguments that also caused the ACS to react, establishing the Division for Industrial & Engineering Chemistry (1908) and a respective journal as recounted above. The shortcomings of mere laboratory chemistry could, according to discursive speakers, be allayed by way of the chemical engineer's combined knowledge of both chemical reactions and mechanical basics needed to scale ounces to tons and flasks to tanks (Sadtler 1908; Teeple 1908; Louis 1917; Carpenter 1917). The third interesting aspect to laboratory scale is timing – these remarks being a reflection of the material world. On the one hand, timing as it relates to shakeups in the organizational field with regard to scientific and technical societies, but on the other hand also with the growth of the number of industrial laboratories in mind. As the discussion above showed, by the 1910s laboratories in corporations had earned a certain acceptance, which is discursively reflected in comments regarding their ubiquity and necessity (cf. Hyde 1909, 907; Little 1916, 78; Clarke 1916, 261). Maybe the laboratory, discursively bolstered by claims of supremacy and rationality, had not produced the grand successes expected by executives and managers (as the histories of the pioneers, particularly the GE laboratory, have shown), and this new notion of scale served as a discursive saving throw to keep laboratories supported by locating failures outside the laboratory? As with the addition of the temporal dimension to the distinction of pure and applied science – with fundamental assuming primacy both in terms of content and sequence – that may have explained why research took so long, laboratory scale could have served a similar discursive function, with the perceived weaknesses of the object being a clear connecting point for various subject positions.

The second change that occurred over the years with regard to laboratory materialization concerns the scope of the laboratory, that is, its tasks and more specifically its types. It must be noted that after the careful introduction and dissection of concepts like pure, applied, and industrial research, so far the laboratory has only been *the laboratory*, and not the industrial laboratory. This happened for a good reason, as it was imperative to characterize the laboratory as object before delving into its specific configurations, similar to science and its variants. Over the entire period analyzed, a plethora of descriptors for laboratories can be found: from technical and university laboratories (Chandler 1900; McMurtrie 1909; Ames 1918) via laboratories typified by scientific discipline (Richards 1903), by sector such as private and governmental laboratories (McMurrich 1907; Hillebrand 1907), and by activity – analysis, testing, control, consulting, and so on ("Research Laboratory at Niagara Falls" 1902; Hillebrand 1907; Mitchell 1912; Mees 1916; Thompson 1917a; Kellogg 1921). One distinguishing factor between laboratories placed in universities and those in industrial corporations was the teaching function as touched upon above. Another one was the secrecy of industrial laboratories being perceived as problematic and manifested in delayed or prohibited publication practices. The need to keep research results private as trade secrets, in order to not lose competitive advantages, was a talking point occurring from 1908 onward and intensifying for the last three events (1912, 1917, 1922). The retention of results was seen as an impediment to the progress of scientific knowledge – and with it, progress and welfare – and as a reason why scientists might struggle in an industrial laboratory (Hillebrand 1907; Burgess 1911; Bogert 1913; Little 1913; Bacon et al. 1917; Mees 1917b). Proper and realistic publication policies were discussed especially during the last two events, in a group of texts that could be considered a new "genre", namely instructions or "how-to"-texts describing proper industrial laboratory setup and administration (e.g. "The New Research Laboratory Building of the American Rolling Mills Co., Middletown, Ohio" 1911; Mees 1916; Clark 1917a; Mees 1917a; Reese 1918), a further indicator of the spread of industrial research laboratories not only in physical objects, but also in taken-for-grantedness as well. Subsequently, this is but one dimension according to which industrial research laboratories were characterized and especially distinguished from their university counterparts.

Another highly interesting attribute of industrial laboratories that developed over time into two distinct types was not in relation to its place vis-à-vis university and corporation, but in the practices it gave room to: research and analysis. The differentiation between analysis and research appeared fragmentary for the early events, as Remsen (1894, 533) illustrates with his distinction between chemical laboratories for those devoted to work in a factory or an analytical laboratory, and those who seek teaching and investigation. By the events of 1908 and onward, authors saw industrial entry-level positions for chemists in analytical laboratories (Burgess 1911), yet also warned of the dangers of only relying on analytical work in laboratories without the application of chemical principles in interpretation, which would result in a "farce" (Brogdon 1912, 685). The lines of demarcation around these notions, indicating what exactly differentiates analytical from research work, are blurry, and from 1912 onward are increasingly tied to the subject positions of research vs. analytical chemists. Similar demarcations were drawn in texts discussing the Research Landscape and laboratory setup, for example, Little listing analytical, testing, and commercial labs while lamenting their lack of appreciation for equipment and organization needed for proper research (Little 1913, 655), Thompson casually mentioning the laboratory – "be it analytical or research" – as the chemist's place in a corporation (Thompson 1917a, 183), Hale (1921, 460) doing the same (cf. also Mees 1916).

What does this tell us? Has the laboratory as an object not explicitly been linked to research as its constitutive practice since it is the place where the scientific method happens? The distinction can be interpreted as a move to position the research laboratory as a distinctly novel part of the organization – a different undertaking from what was done before. This means novel not only in terms of empiricism, rule-of-thumb, and trial and error which were discursive antagonists of the new, scientific ways, but also in organizational terms – as the histories of the industrial research laboratory and the pioneers have shown, departments of testing and control were doubtlessly parts of industrial corporations prior to the first research laboratories, and came to be called laboratories as well. In order to underline why research laboratories mattered, they were constructed differently from mere analytical laboratories, in that they not only broke things down, but also created the new (Brogdon 1912). The opposition echoes in the hierarchy of chemists that will be discussed below.

One last matter in terms of laboratory types may be of interest here – the question of the pioneers. Above, the question was asked of whether the four pioneering laboratories were actual pioneers – as in, the first to build up such facilities – or merely "lighthouses" that were covered favorably and frequently in historic writing. As it turns out, from 1912 onward, the big four's laboratories were recognized discursively as forerunners of research in industry and celebrated for their "brilliant practical and financial success" (Armstrong 1923, 606). GE's laboratory in particular, embodied by Whitney, received repeated praise: "... one of the earliest of its kind in this country, the embodiment of the application of science to industry, has gained a world-wide reputation by the quality of its work and the importance of its results" (Chandler 1921, 160). Kodak's laboratory featured prominently through Mees' writings on laboratory organization and administration (Mees 1916), while AT&T's and DuPont's laboratories were mentioned comparatively less frequently (Comey 1912; Little 1913; Hale 1918; Ogden 1918). Often other firms were recounted as having established similar research endeavors, the National Electric Lamp Association (Hyde 1909; Whitaker 1911) and Westinghouse (Hale 1918) featured prominently, while many others were listed in later texts reporting on the wonders of research (Bancroft 1918, 1922). Of course, such discursive occurrences do not help in answering the question of whether or not the pioneers were actually the first, but they do reinforce the notion of the high visibility of their triumphant achievements from 1912 onward, which may have served as (discursive) prototypes for the establishment of new research laboratories in other manufacturing establishments. Mees' prominent textual production in the discourse of 1917 and 1922, as well as Whitney holding positions that warranted a prominent voice such as the presidencies of ACS (1909) and AES (1911–1912), and recognition through awards such as SoCI's Perkin Medal (in 1921) served to further amplify their message. In this way, the gospel of industrial research was further perpetuated in concrete instances of the object "laboratory".

Subject Positions

After scrutinizing the notions and categories guiding the thought and interpretation of reality, and how they make material reality meaningful in a specific way actors are of interest as a last analytical category. Not just any actors in any capacity, but those assigned special positions in the discourse, termed subject positions. They come into play in two ways: On the one hand, they are constructed as legitimate contributors to a discourse, "allowing agents to participate in a discourse in particular ways" (Munir & Phillips 2005, 1668), in our case those who are allowed to say anything about what science is, how research should be conducted, and where laboratories should be located, to name but a few examples. They feature also in questions of who will be heard discursively and whose constructions of concepts and objects are spread and perpetuated – those coming from legitimate speakers. On the other hand, and this is the component highlighted primarily here, subject positions are constructed as actors – to offer locations for agents to participate in (Maguire & Hardy 2006, 13) there first needs to be a discursive construction of what it means to be a chemist, a scientist, or a chemical engineer. So, instead of focusing on the ways that a subject position offered the means of different levels of participation in the discursive corpus and the respective kinds of texts involved (i.e. who is allowed to say what about something or other), questions were asked about how these subjects were constituted, and how these characterizations changed over time as the results of discursive rifts and events in the field. Instead of analyzing the plethora of subjects mentioned in the discourse – ranging from chemists to managers, and politicians to popularizers – only the chemist and the chemical engineer shall be presented in detail, since the object laboratory clearly showed how it was materialized as a place staffed by chemists. Furthermore, opposition and coalitions toward other relevant groups can be analyzed as part of the attributes of these two subject positions.

Subject 1: The Chemist

What did it mean to be a chemist between 1870 and 1930? What part did the conceptions of science, research, and chemistry play in the attributes constructed of these men of science? Why is the chemist analyzed here, instead of the more general subject position of "scientist"? This is due to the way laboratories were constructed as objects – staffed with men trained in the science of chemistry. In this way, characteristics of the scientist were reflected in how the chemist's identity was constructed. Provisional analysis of the "scientist" showed that the chemist was an instantiation of the scientist, in some cases even going beyond characteristics attributed to the scientist, such as in relation to material things. In order to keep the presentation of the discourse analysis' results as comprehensible and concise as possible, a detailed discussion of the scientist was eschewed here, opting instead to focus on chemists in their function as scientists.

The Chemist as Man of Science

The subject position of the chemist is strongly related to the way that the concepts of science and chemistry were constructed. First of all, the chemist was a man of science, with all of the connotations that followed from association with the concept. Several key features are of note here, beginning with the identification of the chemist as the one who applies the scientific method and uses it in the quest to gather knowledge and ascertain the truth (Carhart 1895; Coulter 1900; Fullerton 1900; Nichols 1918). Due to the nature of their scientific specialty, chemists are similarly positioned vis-à-vis nature, as the concept of chemistry was. Thus, the chemist is the one finding the laws of nature, and "it is the chemist who must discover the truth about the changes that occur or may be made to take place in the composition and constitution of material things" (Scholes 1918, 390). Here, we find again the relation of chemistry to the material world outlined above and the chemist as the one subjugating nature through his mastery of "the atomic structure and the arrangement of matter" (Vaughan 1912, 225). Moreover, emphasis was laid on the chemical education any chemist received, in this way extending the notion of the man of science: Those who apply the scientific method can do so because they were trained at university, in a laboratory, as opposed to tinkering or learning on the job. Ripples emanating in the discourse on chemical education are of course reflected here, for example, in calls for chemical education to be broad instead of specialized ("The Career of a Chemist – A Rejoinder" 1895; McMurrich 1907; Whitney 1909; Messel 1912). The university chemist was of course a teacher, yet many authors underlined that to be a good teacher one would also have to do original investigation – research. Already in 1895 a university instructor was expected to "add something to stock of knowledge by his independent investigation" (Carhart 1895, 395; cf. also Chandler 1900), with investigation in a laboratory seen as a method of fostering independent thinking and creativity instead of memorizing formulas. Interestingly, research by chemists early on was already hardly seen as a solitary activity, rather conducted by a group of investigators. Speaking of the proper endowment of research, Clarke condemned support by paying the salaries of men "working at random", and attacking their small problems in their own ways. Instead, he advocated for "... the establishment of institutions, wherein bodies of trained men should take up, systematically and thoroughly, the problems which are too large for individuals to handle" (Clarke 1902, 136). This notion of research as a cooperative activity gained strength in later years, when more and more laboratories came to be established, and with reference to the formidable successes of German research laboratories (Roeber 1913; Little 1913, 1916). This also signaled a change in the perception of how knowledge and progress was generated – a shift from individual genius toward the collective effort.

In a similar vein is the "Edison method" (Little 1916, 645), lauded in a description of genius and individual achievement by A. D. Little, after which he applauds the "other and even more significant phase of our industrial research, namely, that which involves the coordinated and long-continued effort of many chemists along related lines" (ibid., 648). The chemist was seen as a creative investigator, the one manipulating matter and, at a later point, most effective in a group. As mentioned, a strong discursive shift toward group research occurred over time, and the notion of research gained even further visibility when discussed in relation to the discourse on Preparedness and questions of how the chemist could serve the country best. Unsurprisingly, authors argued for the chemist to be placed in the reserves instead of the frontlines due to their unique ability to undertake research and solve the war's challenges, from dyestuffs to ammunition (Bogert 1917a, 1917b; Stieglitz & Parsons 1917). Another feature of the chemist's subject position that recurred repeatedly was the idea of the chemist as a discoverer or pioneer. While especially during WWI other metaphors surfaced briefly, the idea that chemists as researchers were discoverers akin to pioneers, surveyors, and settlers of the days of the Western frontier remained stable over the whole period (Clayton 1902; Mitchell 1912; Thompson 1913; Parmelee 1921c). Discoveries to be made were likened to "a country rich with undeveloped possibilities" ("Notes and Comments: Room for Research Work in Chemical Engineering", 1908) and a "territory which has not yet been traversed" (Cannon 1911, 66). Such metaphors were also used to delineate differences between pure and industrial researchers, as this pivotal quote by AT&T's Carty shows:

> The investigator in pure science may be likened to the explorer who discovers new continents or islands or hitherto unknown territory. He is continually seeking to extend the boundaries of knowledge. The investigator in industrial research may be compared to the pioneers who survey the newly discovered territory in the endeavor to locate its mineral resources, determine the extent of its forests, and the location of its arable land, and who in other ways precede the settlers and prepare for their occupation of the new country.
>
> (Carty 1916, 514)

Placing the chemist (or the scientist in general) in the context of discovery, within the larger framework of the American frontier experience of less than a decade ago, further elucidates the remarks describing the laboratory as keeping a company ahead – laboratories could serve as the pioneering arm of industry, since they were staffed with pioneers. Furthermore, it serves as another exemplification of how concepts, objects, and subject positions are bolstered by association with other discourses. While not as pronounced as ties to the Conservation Movement or the efficiency craze, "going west" certainly can be understood as a relevant cultural reservoir of ideas during that time (cf. Spero 2014).

But the chemist as university-educated scientist doing pioneering research was not the only attribute that the subject position was constructed with. From 1895 on, and over the whole period, a multitude of authors thought it necessary to underscore important traits of character and attitude, thereby making the chemist special. To cite but a few telling examples, in a reply to an out-of-work chemist in the *Scientific American*, an anonymous author offers a scalding critique, enlisting a variety of people, ranging from Ulysses Grant to successful chemists such as Siliman, Remsen, and Chandler, in order to illustrate what it means to "be a chemist": "There are men, yes, and worse, who have not succeeded in the life work planned for them, not because they were incompetent, but because they were lacking in that peculiar ability of persisting in spite of obstacles that is typical of the best Americans" ("The Career of a Chemist – A Rejoinder" 1895, 211). Next to persistence and grit, truthfulness, honesty, morality, diligence, and accuracy were central characteristics of the scientist, and by extension the chemist (Carhart 1895; Turner 1900; Fullerton 1900; Washburn 1916). The scientific method, then, is used successfully by those with a fitting mental attitude, who are special and stand out from common men (Coulter 1900; McKeen Cattell 1902). Such notions did not lessen over the years, as W. Richardson's list of qualities necessary for being a chemist or chemical engineer shows: "He must have inventive ability, profound knowledge, keen insight, imagination, initiative, tireless energy and that wonderful faculty of elimination of the non-essential" (Richardson 1908b, 809). A certain element of genius certainly played a part in this subject position as well, being an essential element for research (Roeber 1913; Bacon 1917b; Bacon et al. 1917) that needed to be developed in the student (McMurtrie 1907) and was a key factor in the discoveries and developments of the greatest chemists and scientists (Pupin 1895). Though no consensus existed of whether or not great scientists were born or made (McMurrich 1907; Washburn 1916; Weens 1917; Whitney 1909) one feature of this surprising element in the construction of the chemist's subject position needs further discussion: how great scientists and specifically great chemists were described according to the sketched subject position. How was their method, their process described, and what role did genius or anything similar play in it?

A large group of eminent and widely known and appreciated chemists are mentioned. Among them are the (European) forefathers of chemistry such as Lavoisier, Liebig, Pasteur, Faraday, Davy, and Ostwald, and later even American chemists such as Remsen, Whitney, and Langmuir (Little 1913; Reid 1922). Yet several figures stand out. The German chemist Justus von Liebig certainly is one of these figures, and is frequently mentioned in discursive texts (Carhart 1895; Walker 1911; Noyes 1922). Acknowledged as the founder of the first chemical laboratory and the one who transformed university teaching in Germany, these feats were made possible due to his special traits, as constructed in the discourse ("Justus von Liebig and the First Laboratory" 1909). The accomplishments of Louis Pasteur were invoked even more frequently, and, as outlined above, often in relation to the applications of his discoveries to industrial progress, welfare, and wealth (Halstead 1895b; Wiley 1902; Maclaurin et al. 1916; Ogden 1918). Pasteur, a "great savant" (Mason 1895, 851), and one of the "great modern masters of science" (Maclaurin 1916, 43), was acknowledged to have been guided by "higher aims", and it would be from special men like him that a "progressive betterment of human character, the final, most precious result of evolution on this earth" would stem (Halstead 1895b, 612f). Michael Faraday, as one of the founders of the subdiscipline of electrochemistry, was recognized as a genius of special mental facilities: "Unless the scientific investigator has the proper genius for his work, no amount of financial assistance, no apparatus or laboratories, ... will enable such a mind to discover new truths or to inspire others to do so" (Carty 1916, 515). Pupin (1895, 862) describes Faraday as a rare combination of discoverer and philosopher, while Carhart cautions against degrading the genius of Faraday and Hertz by reducing them to the level of the "greatest living inventors" (Carhart 1895, 399).

Changing Conceptions of the Chemist over Time

As we have seen previously, the distinctions of pure and applied according to motive and place played important parts in how the concepts of science, research, and chemistry were constructed. Furthermore, these issues found outlets in the tendency of localization as far as the materialization of science in the laboratory was concerned – the laboratory being a place of science, and its place having a strong influence on objectives. Especially in the industrial laboratory, we already found differences of type, that is, research and analytical laboratories. What about those working in laboratories? Did their subject position change over the years similarly to how laboratory types evolved and how research gathered the additional meaning of industrial research?

Generally, a multitude of descriptors could be found in discursive texts, further denoting the types of chemists. Few trends over time could be identified, yet the overall movement within the discourse is indicative of certain effects that the establishment of industrial research laboratories had regarding attributes of the subject position "chemist". Of the various ways chemists were grouped and typified – ranging from, for example, the opposition of scientific and technical chemists (Duisberg 1896), all the way to research versus service chemists (Russell 1923) – the eventual emergence of the industrial chemist stands out. While in early texts talk focused solely on "chemists", by 1908 authors deemed it increasingly necessary to speak about industrial chemists, first and foremost in relation to the discourse on Chemical Education (Palmer 1908; Walker 1922). The industrial chemist was often also described as a technical chemist (Hillebrand 1907), and descriptions of this type of chemist made it clear that they meant chemists under industrial employ, as opposed to those at a university (Parker 1909; cf. Teeple 1908). Indeed, frequent calls for closer cooperation between these two types of chemists were made, one author going so far as to describe the teaching and research chemist as the "exciter", and the industrial or technical chemist as the "dynamo", with their cooperation bringing "whatever power is to be derived from the science of chemistry, to the industrial world" (Richardson 1908a, 396). That the term "industrial chemist" emerged around 1908 comes as no surprise when one considers the organizational field of academic chemistry, and the changes made to the organization of the ACS' divisional structure in 1908 – traces of which can be found in the discourse. The lack of representation of the technical chemist in the ACS is even lamented (Hillebrand 1907) and explicitly referenced in the founding editorial of JIEC (Parker 1909). As such, the discursive uncertainty of terms can be interpreted as resulting from increasing opportunities for chemists in the chemical industries, and this group of chemists progressively demanding a "voice" among the established subject positions of chemists.

A further distinction within the subject position of chemist that gained a more prominent role on the discursive stage was one already foreshadowed in the ways the laboratory was materialized: research versus analysis. First of all, who the analyst was perceived to be is summarized by Richardson, who wrote that an analyst is a "chemist who, by various devices called methods of analysis, endeavors to ascertain the composition of substances" (Richardson 1908a, 398). Analytical work was generally seen as the basis of all chemical work, being undertaken in industrial establishments, seen as primarily analytical (Richardson 1908a, 1908b; Burgess 1911; Little 1916), and often equated with routine work (Burgess 1911, 617). Indeed, analytical work was termed as less exclusive and pioneering than research, and while all plants would require chemical analysts, not all could do their own research (Richardson 1908b, 809). What is of special interest in analysis as a subtype or attribute of the chemist's subject position is twofold: First, the relation to the research chemist constructed during the later events. In contrast to the analytical chemist doing the routine work of breaking down substances, the creative research chemist was instead busy originating and developing new fields (Whitaker 1911). Over time, a clear hierarchy was constructed, that saw the analytical chemist (or mere "analyst") as inferior to the research chemist, who was schooled in analytical methods but could interpret analytical results and go beyond these in their work (Brogdon 1912; O'Brien 1922). A second interesting facet is how the analytical chemist was described to be perceived by employers. The position of an analyst was often described as an entry-level position for recent university graduates (Burgess 1911; Little 1916).

Subsequently, authors made critical observation about employers' perceptions of analysis as being easy or of low quality (Auchy 1909), and the analytical chemist as being seen as an inferior type (Talbot 1912, 403). The hierarchy of the research and analytical chemist bleeds over into this point, as the lack of appreciation for research was partially attributed to employers confusing both types of chemists (Bacon 1917a), or having bad experiences with analytical chemists pretending to be research chemists (Watkins 1913). Interestingly, while lines of delineation are clearly drawn between the research and analytical chemist as shown here, these two types cannot clearly be linked to university and industrial chemists. On the one hand, analytical work is clearly placed within the confines of the manufacturing plant – the routine work of the university chemist's "technical brother" (Burgess 1911) apparently only the practice of industrial chemistry – and the lamentation of false perceptions often linked to questions regarding the elevation of professional status. On the other, the troubles for research chemists caused by their analytical brethren, and calls for needing to go beyond analysis during the later event-tranches, point toward this type – the research chemist – also having a place in industry.

Opposition and Constituencies over Time: Foremen, Managers, and the Public

The chemists' struggle for legitimacy and recognition occurred not only in relation to their professional standing vis-à-vis the other learned professions of lawyers and physicians, which found its strongest occurrence during early questions of specialization and during the postwar reconfiguration discussing legislative efforts geared toward the chemist's status and the foundation of the AIC, but also in more nuanced ways – which can be highlighted in the discursively constructed ties to other groups.

Surprisingly, the inventor hardly features in the discourse analyzed. Since one of the results of the historical studies on industrial research laboratories was that scientists asserted themselves and their claims to innovation and mastery over inventors, who served as prior sources of corporate innovation, it was expected that such arguments would surface in the events scrutinized. Even though inventors were explicitly targeted in the search heuristic for corpus assembly, only a few texts concern themselves with inventors. If anything, what could be stated is that, regarding the inventor, a certain conceptual or semantic fuzziness between the inventor and the scientist is evident. Especially early on, a variety of ascriptions are made: Inventors are seen as geniuses and pioneers (Thurston 1900; Baekeland 1912), they work in laboratories ("The Status of Electrochemical Industries" 1908), and are born, not made (du Puy 1912), with some accomplished men of science also being described as inventors, most notably Baekeland. For later event-tranches, emphasis was laid on groups of scientists, instead of inventors, being under corporate employ, with growing technological complexity phasing the inventor gradually out (Mees 1916). In this way, discursive arguments were replicated in the explanations of the emergence of industrial laboratories. Only a few texts equate inventors with a lack of scientific training, trial-and-error methods and experience-based expertise ("The Industrial Corporation and the Inventor" 1911; Ames 1918; Armstrong 1923). That is why the first group analyzed in the following is not merely deemed the "group of inventors" that chemists had to assert their discursive dominance over. Instead, true to the interpretive validity of the discursive texts, the analysis proceeded as closely to what was written as possible, and authors spoke of superintendents and rule-of-thumb men much more frequently than of the inventor. Thus, a clear discursive struggle of scientists and inventors[5] could not be identified, instead, tentative links between untrained men and inventors can be asserted.

The first group of interest here are foremen, factory superintendents, and "practical men". Mentions of this group started in 1912 and continued for the later events, though less frequently. With scientific research standing in strict opposition to the old ways of tradition and untheoretical empiricism, the chemists who entered the industrial corporation and their methods were antagonized by factory workers, foremen and superintendents – "... superintendents look upon the chemist as a kind of necessary nuisance whom they tolerate rather than seek" (Brogdon 1912, 685) – with the results of research not appreciated at all: "What a mass of promising research work has been ruthlessly beheaded by conscientious superintendents, and directors in the name of 'practical' results!" (Jacobson 1916, 459). A. D. Little even accused foremen and the like of not having emerged from the "penumbra", resulting in their lack of appreciation and utilization of the research chemist (Little 1913, 653). This group was frequently identified with rule-of-thumb methods and a lack of formal training, and their successes attributed to hands-on experience and years of service (cf. Little 1911; Hadfield 1916; Ames 1918; Noyes 1922). Yet their traditional knowledge came to its limits in a time when efficiency became a growing concern – the description of chemists being received in a hostile way coincides with questions of efficient production, waste management, and product innovation, outdated and inefficient ways that only persisted due to the old guard (cf. Maclaurin et al. 1916, 65). Soon after the concerns over efficiency, preparedness and war became the watchwords connected to the conflict between the old methods and the new, scientific knowledge and their respective practitioners. Often, the relation of the practical, untrained man to the university-educated scientist was seen as a linear displacement: "The day of the typical 'practical man' is passing and there is dawning the day of the efficient scientific man" (Washburn

1916, 1150; cf. also Roeber 1912; Clarke 1916). Even after the war, complaints about ignorant foremen continued: "Far too many of our plant and department superintendents do not yet comprehend how the prosperity of their companies can be augmented by a greater use of scientifically trained men – men from the technical high and manual training schools as well as from the universities" (MacDowell 1922, 431). To summarize, the distinction found in the concepts of science/research as different and better – more efficient, better for the war effort, and the solution for the postwar economy – can also be found in the subject position of the chemist as the antagonistic relation toward them from a group best condensed as "practical men". Why did this oppositional relationship only occur from 1912 onward, if chemists were widespread in industry years before? One possible interpretation runs parallel to the introduction of laboratory scale – chemists in their laboratories lagged behind in producing their promised results, leading to the construction of an opposition group that discarded their knowledge of underlying principles and their improvements to tried and tested processes. Another perspective takes into account how research laboratories gradually became established as a new and distinct part of the industrial corporation that was not troubled with the daily, routine concerns of the shop floor. Due to this distance, ideas from the research laboratory may not have met manufacturing reality (university chemists' lack of practical experience being a frequent concern within the discourse on chemical education and a connecting point to the subject position of the chemical engineer), instead meeting opposition. One last topical interpretation lies in the chemist (and especially the chemical engineer) being in the position of efficiency expert, with the old ways failing in the race for greater efficiency, and the development of the scientific method, which was employed by chemists in the laboratory as an answer for these troubles. Thus their natural enemy in achieving greater efficiency were foremen and the like.

The second group is a combination of actors in leading positions in the industrial corporation, variously called businessmen, executives, or managers, and those with the means to finance research ventures: capitalists and bankers. In contrast to the way that discursive speakers characterized foremen and superintendents, the relation to this group was denoted as one lacking in appreciation, mainly due to absent knowledge of who the chemist is and what their laboratory research could accomplish. Tenuous relations to this group were mentioned starting in 1912, with the discussion gaining speed and visibility from 1917 onward. A lack of appreciation was often identified with inadequate compensation, and many authors lamented little pay for chemists when the worth of the services of lawyers – as recognized members of a learned profession – was evident, appreciated, and thus appropriately compensated for by executives (Watkins 1913; Stieglitz 1917a; Burgess 1923). The source of the perceived erroneous valuation lay in false perceptions of chemistry which at times was regarded as either simple, the chemist being only a day laborer (Stewart 1912; Watkins 1913), or too impractical, the chemist coming from university halls being perceived as an "impractical man" (Hemingway 1917, 354; cf. Thompson 1918). Interestingly, these lamentations were often made with respect to a lack of appreciation for *research* chemists, with manufacturers being blind to the value of doing organized research, whereas the services of chemical analysts were commonly acknowledged (and regarded as simple and routine) (Watkins 1913). Generally, the missing knowledge of what research is and what it can do was supported by the notion of the special mind of the chemist – "The mind that finds the way to do new and improved things must see visions that are entirely obscured to the minds that finance, direct, manufacture and market" (Choate 1917, 244) – and chemistry and research proving to be a mysterious province for many executives and bankers (Matthews 1916; Little & O'Reilly 1917; Roeber 1917). The lack of knowledge about research and those doing it was connected to the efforts of Popularization after WWI, finding expression primarily in appeals to learn how to "sell" chemical research: "Selling research consists in convincing all with whom the research worker has to do, that his research is worth while doing, and that the results will show a gain in quality or cost or performance over the methods with which the user is familiar" (Skinner 1922, 168). Due to the inherent foreignness of the laboratory and chemists, with executives and bankers essentially speaking different languages, the chemists would need to learn to translate laboratory results into a language that corporate decision-makers could understand (Clark 1917b; Teeple 1922; Hyde 1922; Burgess 1923). A campaign to "evangelize" the business world (Hyde 1922, 291) was seen as the solution to inadequate salaries, professional compensation, and the closing down of research laboratories after the war (Bolling & Maze 1922; Eisenschiml 1922c). This also connects to calls for leaving the confines of the laboratory, becoming active in other corporate functions, and rising to the executive level, as outlined above. In summary, the difference is striking in the relations constructed between chemists and the first group, foremen, and this mélange of executives, managers, and capitalists as the second group. Instead of interpreting the lack of status and compensation as hostility, it was merely put down to different languages and the ignorance of many, while few manufacturers had already seen the light (Stewart 1912). Moreover, while the time of the "practical man" was deemed over, in turn heralding the chemical age, it appears that for discursive authors it was obvious that executives were here to stay and needed to be swayed in their support for research laboratories, with the idea of putting chemists in other positions than the laboratory only gaining traction during the last event-tranche analyzed, 1921–1923.

What can we learn from this in an attempt to map the changes to the subject position over the years? For one, it further illustrates how chemistry was materialized in the laboratory, and how the subject position of the research chemist was associated with it – chemists were now asked to move out of the laboratories, branching out and spreading knowledge of chemical wonders (Bolling & Maze 1922). It also aids in understanding how the discursive level probed here is interconnected with the organizational field and the events in the field: Professionalization and popularization are reflected in the establishment of, for example, the AIC and popular campaigns already discussed, with our knowledge of these happenings now supplemented by a thorough understanding of the way that the relations of the subject position of the chemist were constituted and how they were transformed.

Subject 2: The Chemical Engineer

The second subject position to be outlined here is, of course, a minor one compared to the (research) chemist, yet still relevant by nature of the material analyzed itself: Discussions of chemical engineering came up frequently, and ignoring them would render any meaningful analysis of the discourse only fragmentary. Moreover, such discussions informed the way that science was seen to be applied in industry – through the materialization and location of the industrial laboratory – while also sharpening our understanding of the chemist's profile. Thus, in the following, two themes shall be highlighted: On the one hand how the construction of the "chemical engineer" was undertaken and how it relates back to events in the organizational field. And on the other, in what way the Conservation Movement, the efficiency craze, and WWI transformed the subject position, in turn repositioning the laboratory vis-à-vis corporations and the shop floor.

Who Is the Chemical Engineer? Between Engineering and Chemistry

Discussion about this subject position started with the earliest event, 1895, and intensified in the following years, peaking around 1908–1912, when it was primarily discussed by the chemical and industrial speaker groups. But with each event, the focus of the discussion seemingly shifted. Around 1895, the chemical engineer – a term not yet coined – was discussed in relation to necessity and demand, with various authors calling for an industrial worker that would combine a knowledge of chemistry and of engineering in one person, as both specialties were acknowledged as necessary and foundational to modern industry: "In connection with the working force of the German color factories … that experience has led directors to employ educated engineers alongside the research chemists and so to recognize the fact that engineering capacity is necessary to the practical and industrial application of chemical reactions" (McMurtrie 1895, 293; cf. Platt 1894; Chandler 1900). The term "chemical engineer" was coined around that time, and the focus of discursive texts moved toward the question of where on a scale between chemistry and engineering the new subject position should be placed. While many authors emphasized the ability to combine both sides equally (e.g. "A Proposed Society of Chemical Engineers" 1907; Parker 1909; "Industrial Research" 1911; Thompson 1917b), general consensus claimed the chemical engineer to be a chemist first, and an engineer only in a secondary capacity:

> In my opinion, a chemical engineer is, primarily, a chemist, but essentially a man of affairs and executive ability who is engaged in the application of the principles of chemistry in the arts and industries, with special reference to the commercial development of the industries and establishment of new ones, to perfecting the arts, and by his advice and cooperation, assisting other engineers in the practise of their professions.
>
> (Richardson 1908, 82)

The necessity of proper chemical training for the chemical engineer, instead of merely hands-on chemical knowledge picked up on the job, was stressed (Richardson 1908a, 401; cf. "Why Not 'The American Society of Chemical Engineers?'" 1907; Sadtler 1908; Palmer 1908). In this way, the subject position of the chemical engineer was clearly and distinctly constructed as one belonging to the domain of chemistry, and not represented by engineering societies. To refer these findings back to the organizational field, it is highly illustrative how questions of professional identity were especially prominent prior to and during the establishment of AIChE (1908), as also illustrated by the discursive texts cited above: Almost all of them were found in the event-tranche 1907–1909. These questions' pertinence is framed by the boundary-work necessary to establish the AIChE, which was explicitly positioned not as a rival organization to the ACS, but as an organization with a narrower scope, membership, and mission, as also reflected in discursive texts debating whether such an organization should be established ("Why Not 'The American Society of Chemical Engineers?'" 1907; "A Proposed Society of Chemical Engineers" 1907; "Some Thoughts on the Organization of the 'American Institute of Chemical Engineers'" 1908; "The AIChE" 1908). The need for the specialization and professionalization of chemical engineering was often supported with reference to the successes of the German chemical industry (McMurtrie 1895; Sadtler 1908; "Notes and Comments. The German Chemical Industry" 1908; Palmer 1909), further illustrating how concepts – the German chemical supremacy – were utilized in discursive constructions and as legitimations, in this instance to bolster the case for an organizational home of the chemical engineer.

Several attributes are of note in the subject position of the chemical engineer, that underline its relevance to the industrial laboratory. First of all, many of the texts outlining who the chemist is supposed to be often also talk about the chemical engineer, extending categories and attributes – mastery over nature, rationale and training, the use of the scientific method, higher education, etc. – toward those with chemical training, be they chemists or chemical engineers, as is evident in a variety of texts on chemical engineering education (McMurtrie 1901; Richardson 1908). The chemists' struggles for public appreciation were mirrored in those of the chemical engineer (Matthews 1916; Withrow 1916), and untrained workers or business men were occasionally invoked as oppositional groups (e.g. "Engineers, Practical Men, and Theories" 1913; Thompson 1917a). But whereas the chemist's subject position sprouted a variety of types – from the analytical chemist, to the industrial chemist, and to a pure research chemist at a university – the chemical engineer was always distinctly placed in an industrial context. The chemical engineer's tasks were seen as ranging from overseeing production work to interpreting results, in turn altering and improving processes and machinery on the shop floor (Platt 1894; Palmer 1908), while also applying the principles of chemistry in an industrial context (Richardson 1908; Richardson 1908; Booth 1912; Whitaker 1911, 1912b; Bacon 1917a). This positioning came to be important once the shift to matters of laboratory scale happened. Furthermore, strict lines of delineation were drawn between the chemical engineer and both the analytical and industrial chemist during the formative years of the AIChE (Richardson 1908, 83).

The chemical engineer's field of activity was constructed as being much broader than that of the industrial chemist – who stayed confined to chemical manufacturing (Roeber 1908, 309) – and the chemical engineer's methods were going farther than those of the analyst, due to a close acquaintance with the problems of chemical manufacturing (Sadtler 1908, 35). Such descriptions could also be found in the formative texts of the AIChE (cf. "The AIChE" 1908). But, while more than an analyst, the chemical engineer was also clearly not a *research* chemist. The chemical engineer would utilize chemical knowledge but in a different way – in a diffuse space between scientific chemistry and its applications to machinery and production processes. Even though the improvement of processes was seen as an essential part of the chemical engineer's work, laboratory research was deemed the research chemist's domain (cf. Richardson 1908a; Bacon 1917a), while the chemical engineer applied the outputs of the industrial research laboratory to manufacturing (Whitaker 1911, 9). Thus, the emerging subject position scrutinized here is located not only between the domains of chemistry and engineering but also between the research laboratory and the shop floor. In this way, the chemical engineer serves to further inform our analysis of the discourse surrounding industrial research laboratories, especially their materialization and struggles as an organizational entity, which is best illustrated by the quest for efficiency and matters of scale that will be discussed next.

From Conservation to Efficiency and the Laboratory

As already touched upon in the discussion of the Conservation Movement and the efficiency craze, connections to other, adjacent discourses played a major role in the construction of the chemical engineer's subject position. They illustrate, especially from 1908 onward, how discursive shifts influenced and were used to modify existing subject positions, and by extension concepts, in this case the concept of chemical engineering (which was of lesser interest to our inquiry on industrial laboratories than those *doing* the chemical engineering, hence only the chemical engineer being discussed in detail here). Once the term "chemical engineer" was coined and its meaning – a chemist with engineering knowledge – was discursively fixed, authors started associating the subject position with the Conservation Movement that was just about to gain momentum. These associations primarily happened through references to Roosevelt's address for his appointment of the Commission for the Conservation of National Resources. The Commission's mission as outlined by Roosevelt was seen as "almost a good definition of the work of Chemical Engineers" ("The AIChE" 1908, 5; cf. Roeber 1908; Richardson 1908; Bailey 1911; Benner 1912). In this way, the chemical engineer gained an important area of responsibility in addition to aiding industry in best utilizing its resources (Sadtler 1909, 107).

This extension of the subject position, which was also heavily used in arguing for a professional and organizational home that ultimately came to be the AIChE, was only possible due to the chemical engineer's position between chemistry and industry, and between the research chemist and the analyst. With the Conservation Movement's thrust waning by 1912, these new attributes as claims to the domain of chemical engineering and the skills of its practitioners could readily be applied to the need for efficiency: "Efficient production and the economic management of our manufacturing plants are essential features to our commercial development, and it is in this field that the greatest results are to be attained in the conservation of our natural resources" (Whitaker 1911, 9). The focus on production and business efficiency was already outlined above, as well as how science and research formed essential elements in the discursive arguments over how to improve efficiency. Subject positions played an important part here as well, as the trained man using the scientific method was distinctly positioned as an efficiency saver – either by doing research in a laboratory in the case of the (research) chemist, or by aiding in bringing together the different realities of laboratory chemistry and industrial production (cf. Booth 1912; Brogdon 1912). Also, it is exactly here that we find the issues of laboratory scale again. Discursively, issues of scale served to reposition the laboratory vis-à-vis other parts of the company, while also supplying a role for those who did the material translating, the chemical engineers. Material translation as an accepted part of the subject position was only possible through the unique combination of chemical and engineering knowledge that every chemical engineer was supposed to have. It was this second part, engineering, that played a pivotal role in arguments found in discursive texts. Teeple (1908), for example, claimed that the university training of chemists could not cover everything, and hence that they had a lack of familiarity with factory apparatuses, materials of construction, and common chemicals used – an area of knowledge imparted to the chemical engineer as part of their engineering training. Similar arguments can be found in Sadtler (1908) and Parker (1909), who attributed laboratory failure to the research chemists' lack of knowledge about machinery, whereas Louis (1917) and Carpenter (1917) explicitly point toward a need for engineering knowledge. Then again, authors are careful to not concede this domain to the engineer, by reiterating how it is still *chemical* processes being moved from laboratory to factory ("Where Are the Leaders?" 1918), and how essential fundamental chemical knowledge is to this task:

> The chemical engineer is sorely needed at this point to take chemical principles and engineer them just as the mechanical engineer engineers the physics of heat, or the electrical engineer engineers the physics of electricity. ... Furthermore, our researches, where industrial application is sought, must be transferred from the beaker to the tank, from the funnel to the filter press, from the evaporating dish to the vacuum pan, from the distilling flask to the still, and so on, ...
>
> (Whitaker 1912a, 154)

Issues of laboratory efficacy and credibility converged with the emerging professional identity of the chemical engineer, two discursive strands harnessed especially by speakers surrounding chemical engineering, that is, published in Chemical Engineer, Electrochemical Industry, and the Transactions of the AIChE. This case can be regarded as the clearest example of special interest groups forming and utilizing positions of voice to get their message spread, in contrast to many of the other facets of the discourse – be they conceptual transformations or the specific materialization of the laboratory at various places – where interpretations held much more stable over different groups, often by virtue of many authors publishing in a variety of journals, and influential texts – such as presidential addresses and controversial editorials – being referred to by many others. Only when accounting for this organizational field-level context and the history of the various groups involved, can quotes such as "The manufacture of chemical products on a small scale in a laboratory is a relatively simple operation, but when this manufacture is conducted on a large scale, involving the investment of considerable capital, then the problem is much more complicated and much more difficult to solve" (Thompson 1917b, 1007) be accurately classified – in this case as a eulogy on chemical engineering by the president of AIChE himself.

To sum up these reflections on the subject position, we not only add another piece to the discursive puzzle surrounding the industrial research laboratory, but also see how other discourses and events influenced discursive constructions and further our understanding of the process by which discourse feeds off and in turn becomes perceived reality. Moreover, only by carefully reconstructing subject positions – the scientist and the chemist – and staying alert and open to additional features of discourse occurring in the analysis – the importance of the chemical engineer – can explanations of the industrial laboratory based on the scientist's claim to superiority be probed for their accuracy. As it turns out, scientists did assert their mastery of nature and rational methods as opposed to those untrained, but not without also laying claim to being geniuses and having special minds.

Summarizing Results

Let me now summarize the results of the discourse analysis chronologically and with a special focus on the interrelations constructed between the categories: how concepts, objects, and subject positions not only changed over time but also vis-à-vis each other.

In 1895, Chemical Education and Specialization were topics of frequent discussion. The question of how to best prepare students of chemistry at universities and colleges for a future of unclear places of employment – be they university, private laboratories, or within corporations – evoked the expression of many different ideas. Similarly, the increasing specialization of the scientist and the chemist in particular, toward a highly-trained specialist who would not be able to understand current developments in other sciences or even other branches of chemistry, was frequently met with concern or indifference in the name of progress. In both these discourses, ideas about what science and chemistry meant at the time could be found. Science was a quest for truth, leading to superiority over nature and material welfare, by applying what could be recognized as the discursive "core" of this concept: the scientific method. Chemistry, as a science, boasted similar attributes along with a special relationship to the material world, as it allowed the transformation of matter which lead to utility in industrial applications. Science and chemistry were materialized in the object "laboratory", as the physical manifestation of science, thus the laboratory became the place of the scientific method, with chemistry's subjugation of nature only becoming possible within the walls of a laboratory. The laboratory, staffed by professional scientists, allowed for both teaching and research. The subject position of interest at this early event was that of the chemist. The chemist, as a university-educated man of science, came to be regarded as the applicator of the scientific method – imbued with honesty, truthfulness, and accuracy and in line with the forefathers of chemistry who were often described as geniuses, or of special character at least. Thus, for the event of 1895, we find the expressions of that time of what science meant to discursive speakers, what a laboratory was to those who talked about it, and who would be employed in a laboratory. Of course, many of these ideas would go on to be challenged and translated over the years, but especially the clear, unquestionable construction of the laboratory as *the* place of application of science – instead of, say, the university – proved to be key in understanding why, when industrial leaders brought science within their walls, they did so by setting up laboratories.

1900 was the year of the foundation of the GE Laboratory. The discourse on Chemical Education still proved to be of concern for many speakers, yet neither the concept of science nor chemistry changed in its discursive construction, remaining a quest for truth embodied in the scientific method. Though the laboratory, as science's materialization, became increasingly codified as the place where not only men of science were employed, but also chemists worked; thus chemists were clearly discursively located within the confines of a laboratory, wherever said laboratory might be located – at a university, within a corporation, as part of a foundation, etc. The chemists themselves increasingly came to be depicted as pioneers, surveyors, and frontiersmen, pushing ahead into the unknown. Interestingly, around 1900, another subject position entered the discourse: the chemical engineer. Still diffuse in its boundaries, oscillating between engineering and chemistry, the chemical engineer's domain, place of employ, and education came to be increasingly discussed. In the end, 1900 did not bring a revolution in how concepts and especially the object of laboratory were discursively constructed, and the newly minted GE laboratory was neither heralded as a the corporate revolution it is at times seen as today, nor frequently discussed at all.

The next event, in the year 1902, saw the establishment of DuPont's two laboratories: the Eastern Laboratory and the Experimental Station. As a newly adjacent discourse, the Research Landscape – meaning how to fund the quest for science, and where that quest should be located – started to be of concern. In these discussions, it became obvious how the concept of science essentially functioned as a synonym for the concept of research, which started to be discursively named and used more and more from 1902 onward. Research, then, meant the practice of doing science, which was characterized as slow, methodical, and organized, while also leading toward superiority over nature, progress, and material wealth. Often, research was described as belonging to a historical trajectory, as the logical step to surpassing the "old ways" of trial-and-error and rule of thumb empirical investigations – the organized, methodical manner of research being obviously more powerful and appropriate for the times. The object of laboratory as the place of research was often linked to the discourse on the research landscape. Here again we can see the discursive function of localizing: Research was placed within the laboratory walls, while the laboratory itself kept its dual function as the place of teaching and research. Furthermore, these calls were underscored by the utility – that is, industrial applicability – of laboratory findings. The construction of the chemist, ever present as the person working in the laboratory, did not change for this event. The discussion of whether chemical engineers were chemists or engineers intensified, though, and was connected to the discourse on specialization – "mere" chemical knowledge not being enough anymore, and the specialization of education being necessary to imbue chemical engineers with knowledge of machinery and engineering basics.

In 1908, the year that saw the foundation of the ACS Division of Industrial and Engineering Chemistry, specialization was still a relevant discourse. This time, though, it was mainly discussed with regard to organizational specialization – in ACS subdivisions or dedicated societies – instead of in a disciplinary context, that is, pertaining to the various branches of chemistry. With the Progressive Era in full swing, both the Conservation Movement, and with it corporate quests for efficiency, were regularly occurring themes. These new discourses saw shifts in the construction of concepts and objects. Especially the concept of chemistry was now positioned as the solution to efficiency troubles and issues of conservation, in contrast to traditional and failing rule-of-thumb methods. Furthermore, within chemistry the distinction between analytical and research chemistry grew stronger, paralleling an increasing differentiation of laboratory types, such as analytical, testing, or research laboratory, and types of chemists, further illustrating how the three analytical categories of concepts, objects, and subject positions are interrelated. Needless to say, a similar relation regarding efficiency was constructed discursively for both the laboratory and chemistry. In 1908 for the first time the term "industrial chemist" emerges, a role which was seen primarily as a technical chemist. An end point was reached in the discussion of the education and proper location of the chemical engineer, with those claiming chemical engineers to be chemists first taking a semantic victory by associating them with attributes established for the chemist's subject position. The chemical engineer, then, would apply chemistry in an industrial context, making them prime actors in conservation.

The next event, 1912, saw the establishment of the last of the pioneer laboratories: Eastman Kodak. In addition to the Conservation Movement and increases in efficiency, the discourse on how to fund research came to be discussed again, as well as legislative efforts surrounding the chemical industries. In the realm of concepts, "industrial research" came to be a widely used term, associated with the same attributes as the more general term "research". It was also constructed as a necessary foundation for industrial progress, and was strongly linked to applied chemistry. The occurrence of the term illustrates not only how it is discourse shaping the organizational field, but also how it in turn shapes discourse, as the emergence of a concept discursively delineating the practice of research in industry shows. By 1912, having a laboratory was seen as a necessity in industrial organizations, either as an analytical or a research laboratory. Interestingly, while no clear localization of analytical and research chemists happened in the discourse – they could work in industry, at university, or elsewhere – a clear hierarchy emerged, with research chemists seen as being superior to "mere" analytical chemists. Yet, by their association with the concept of chemistry, chemists as a group were positioned in clear opposition to foremen and other traditionalists who favored empiricism and rule-of-thumb methods, a clear discursive move to associate chemistry and chemists with the quest for efficiency – which could only be guaranteed by the use of rational, planned, scientific methods, and not by clinging to the old ways, as chemists accused their opposition of doing. Paralleling the distinction of research and industrial research, the concept of chemistry split increasingly into pure and applied chemistry, the location of which proved to be an interesting result. Localization happened according to motive, and not the actual practices involved. Thus, pure chemistry searching for truth happened within a university, whereas its applications found a home in corporations.

1917 saw the United States enter WWI, which resonated widely on the discursive level. Topics such as preparedness for war, legislation to shelter American chemical industries, as well as the popularization of the achievements of science and the role scientists and chemists played in the successful war effort were regular occurrences. Unsurprisingly, this also resonated in the way that concepts were constructed. Industrial research (and chemistry) was now seen as a prime factor in preparedness, and to catch up with Germany, one would have to follow their example of intertwining industry and chemical research, as well as the system of German higher education. Many authors discussed not only how to popularize the triumphs of science, but also how to sell industrial research to those managers and executives not yet convinced of its powers – a task greatly aided by the associations made with the war effort, that is, industrial research being essential for victory. The laboratory became the place where said preparedness was "made". The difference of scale between laboratories and factories that came to the fore around 1908 still played a role, now concerned with the mammoth task of supplying an industry geared for war. Hence, discussions of proper laboratory setup and equipment intensified, while differences of scale were proposed to be bridged via different stages of research. The discourse on popularization distinctly resonated with the chemist's subject position. Instead of progress-averse foremen and inventors, managers and bankers were now in opposition due to a lack of appreciation of the services of the chemist. Thus chemists were called to move out of the laboratory, their localization now expanding, and toward executive functions.

The last event, 1922, was chosen to assess the post-war return to normalcy, which also saw the establishment of DIR, a meeting group of industrial research directors. With only a small number of texts available, few discursive shifts could be gleaned, but new and continuing discourses slightly modified existing concepts and subject positions. Several discourses returned to the discussion, such as the Research Landscape – how to organize a national system of research in the United States post-WWI; as well as legislation, how to keep the US chemical industry ahead of the possibly soon to be resurgent German one; chemical education; and the return to Postwar Normalcy, along with the structure of the organizational field having changed and new networks having been spanned by new organizations founded in the war. Lastly, concerns of greater efficiency in all walks of life as well as the popularization of science appeared as related discourses linked to this discourse. The concept of chemistry was strongly connected to these discourses, especially popularization and legislation, since they directly concerned the perceived powers of chemistry. The successful war effort had proven the value of chemical (and industrial) research in the eyes of many speakers, however the public was not yet aware of the many ways that chemistry touched their lives. Popularization was deemed an effective way for enhancing the chemists' standing, as well as for convincing political actors of the need for safeguarding all of the new branches of chemical industry that arose during the war.

In this summary, the way that concepts, objects, and subject positions were constructed and changed was in prime focus. The discourse analysis clearly illustrates how other discourses – be they field-internal (such as Research Landscape) or society-spanning (e.g. Conservation or Efficiency), inspired by external events (e.g. the beginning of WWI) or by events on the field level (e.g. discussions on specialization and chemical education brought about by the professionalization and differentiation of chemistry) – were picked up discursively and subsequently influenced the construction and translation of concepts, which is one of the prime tenets formulated in the discursive perspective on the diffusion of ideas: Ideas need to build bridges to larger discourses and happenings in the field to become and stay relevant. Furthermore, the minute and precise ways that concepts, objects, and subject positions were constructed and resonated with each other shows how the level of discourse and that of the organizational field relate to and feed off of each other. By locating science not at a university or in a genius scientist's head, but within a laboratory that could be established anywhere, moving research into industry became possible on a discursive level. Moreover, once the practice of establishing industrial laboratories had spread in the 1910s the discursive concept of *industrial* research emerged, to make sense of the new reality in the organizational field – industrial research that needed to be delineated from the other ways of doing science, placed in temporal succession to its pure variant, and so on.

Notes

1 It needs to be remembered that the unique organizational field used as an analytical reference point to find access to this specific discourse includes a variety of layers of larger discourses (business, governmental, etc.) that cannot all be accounted for in the contextual analysis, but which finds entrance through the discursive corpus and will play a part in the analysis (see especially the beginning of this chapter). In this way, relevant texts codifying, for example, prevalent conceptions of control (Fligstein 1990) may serve as central, legitimating texts for a business organization, while playing no part at all for an academic organization and its discursive background – hence the heterogeneous set of references.

2 Due to the small number of texts for the first event (1895), it is possible that science and research were already used synonymously, yet simply not covered in the small sample. For the next two tranches, 1900/1902, it became clear that both terms seemed to have been used interchangeably.

3 The craze for efficiency did not stop at the walls of the laboratory, with administration and apparatus being a point of contention now – adequate leadership and equipment were seen as a means of raising laboratory efficiency (Whitaker 1911; Bacon et al. 1917).

4 This bold proclamation of the failures of research laboratories can be found in an editorial on the first pages of the first issue of the ACS' "Journal of Industrial and Engineering Chemistry", outlining the necessity of a specialized journal for technical and industrial chemists.

5 It needs to be noted, however, that inventors have no voice here – since the discourse analysis focused specifically on discourse surrounding chemists and scientists, possible speakers for inventors were not included in the field. Of course, some of their triumphs were celebrated in the popular press, giving us a glimpse through the pages of *Scientific American* and *Popular Science Monthly*.

7 Discussion and Conclusions

Summary

In this last chapter, the ramifications of the discourse analysis' results will be discussed: How can the results be tied back to the highly detailed knowledge of the field found in Chapter 4, and how can the results be assessed in light of said knowledge? What are the ramifications of the discourse analysis for the established history of R&D laboratories as retold in Chapter 3? And lastly, how do the results fit with and potentially modify our theoretical understanding of how ideas are translated and connected to grander cultural schemata in order to spread successfully?

The Industrial Research Laboratory as Seen through Discourse

What do we know about the emergence and spread of the industrial research laboratory, and what do we know *now*, having been enriched by reframing the story through the lens of diffusion analysis? As historical scholarship showed in convincing detail, around 1900 American corporations in the electrical and chemical industries began to establish in-house research laboratories as a fixed and stable entity of the organization, and staffed by university-trained scientists. Among those corporations, several stood out for various reasons, with their famous laboratories receiving intense scholarly scrutiny: GE, Du Pont, AT&T, and Eastman Kodak. As their individual histories show, these pioneers established their laboratories in different ways, at different times, and for different reasons. The labs were established at differing parts within the organizational structure, sometimes built from the ground up, sometimes supplanting already-existing testing or development setups. The precise timing of when organized research started at these four companies varied by more than a decade. In addition, the efforts made by actors within the firms, pushing for the establishment of laboratory facilities, illustrate different reasons – from expiring patents and intensifying competition to what may be called fashion, while also taking different paths. At GE, a coalition of various actors enacted a bottom-up push for laboratory research, while at Kodak Eastman himself decided in favor of research. Du Pont's unique configuration of two laboratories is due to rivaling groups within the company fighting for hegemony. It is evident that the whole process of the R&D laboratory spreading throughout American industry can be regarded as one of diffusion. But these four cases illustrate how "the research laboratory" as an organizational entity was made to fit local variations in individual acts of translation-in-action, reshaping the laboratory, its mission, and practices according to local requirements. Understanding diffusion as translation, what spread here is the "form" of doing organized research in a laboratory setting, whereas the "content" of what actually happened in those laboratories varied considerably.

Even though reliable statistics are hard to come by, surveys such as those published by the NRC show that the industrial research laboratory diffused successfully from 1900 onward, indicating that by the end of WWI the research laboratory had become an expected, normalized part of the modern corporation. Now, it would greatly oversimplify the diffusion of the industrial research laboratory by assuming that many firms set up laboratories at roughly the same time and only by way of the efforts of individuals – especially since the detailed case studies of the pioneers and other firms show how they were not inspired by each other. Thus, the spread of R&D cannot be regarded as a case of simple mimesis. The hunt for conditions mutually shared by these companies points to firm-level and environmental factors playing a part, such as firm size, sector and assets, as well as patent and antitrust legislation influencing corporate strategy, and an increasing number of university-trained scientists moving out of a growing academic system. As Sturchio (1981, 85f) notes, it was certainly a "combination of these factors – economic, institutional, technical and intellectual" that brought about the birth of the industrial research laboratory and the eventual shape it took in the giant corporations of the early twentieth century.

However, these factors on their own can neither explain the shift to laboratory-based research carried out by scientists, nor the successful diffusion and translation of the laboratory. Existing explanations – based on business strategy that was geared to fend off competitive threats, or scientist-entrepreneurs within the corporations putting their ideas into practice – mostly fail to successfully account for all these layers combined *and* what I call the cultural layer. As research in the tradition of the diffusion of innovations shows, the cultural attributes of a social system and the environment wherein a new idea is spreading play a pivotal part in successful diffusion. Thus, while past research informs us that the R&D laboratory certainly fit within business strategies of the time that stipulated pulling risk inside the firm, and within a legal environment of antitrust and patent legislation being enforced even more strongly, what my analysis adds is an analysis of the cultural attributes of these corporations' environments. The idea of the culture surrounding science and the scientist figuring greatly into the diffusion of the industrial laboratory is not only informed by diffusion research emphasizing the necessity of cultural fit or appropriateness but also by historical scholarship indicating a *cultural shift* from inventors to scientists as the locus of truth, progress, and expertise in the public perception. Hence, science became important beyond the mere manpower supplied by the US academic system gaining strength and professional identity with the outgoing nineteenth century. The meanings ascribed to science and the scientist formed another layer in the puzzle of understanding how the industrial research laboratory went from being a novelty in 1900 to an institutionalized part of the modern corporation by the 1930s.

My analysis adds to the characterization of a corporation's environment by understanding it as an organizational field, which offers the means of drawing boundaries around those companies establishing laboratories, without resorting to simplifying designations such as sector or industry. Due to the detailed inspection of the field of US chemistry from 1870 to 1930, several important puzzle pieces in the R&D lab's early history could be found. First, a "laboratory" was not a radically new concept to many companies, since testing and control laboratories had long been in use, as well as a few independent laboratories sometimes specializing in product development. Instead, the practice of placing a laboratory *inside* the company was new, and these laboratories doing some kind of *research* was a novelty as well.

Second, US academia as a whole, and especially the discipline of chemistry, expanded and professionalized considerably in the nineteenth century's last decades, as several proxies – such as membership in the ACS, the growth in number of scientific societies or degree conferrals in chemistry – show. But these indicators not only show the growth of academic chemistry in numbers but also hint at the ongoing process of the professionalization of the academic chemist.

Third, when shifting focuses from academic to industrial organizations in the field, similar growth trends are visible. The industry as a whole expanded strongly from 1900 to 1925, which is mirrored in the growth of professional associations and other organizations for chemists, who formed the largest share of the staff of research laboratories in their first decades. Chemists thus gathered not only visibility in industry but also representation through professional organizations. The detailed analysis of the organizational field provides the groundwork for accessing the cultural layer surrounding the early R&D laboratories not only by properly charting the environment of the pioneers but also offering avenues for finding speakers, as producers of units of discourse, that need to be mapped in order to account for the cultural shift surrounding the locus of expertise, which is part of the cultural attributes of the organizational field in question.

Organizational institutionalism illustrates how a field's institutions – going beyond rules and norms to the very definitions of concepts and categories that make up social reality – condition organizational behavior, such as the decision to establish a laboratory. It happens in the first place by supplying notions of what a laboratory is and what it can do, who works in it, and so on, to the players in the organizational field of corporate R&D. Furthermore, due to field-level dynamics, organizations are subject to isomorphic pressures in their quest to make not only rational but also legitimate decisions. Institutional isomorphism manifests concretely in the mimesis of other organizations that are perceived as successful and in professional standard-setting. Hence, to understand and properly model the cultural layer that plays such a large role in the successful diffusion of the laboratory, and the way it was translated within a variety of firms, the makeup of the institutions in the field that was staked out needs to be assessed. For this, the analytical notion of discourse was introduced. Institutions are then understood as discourses – textual entities that construct reality. Discourse and the discourse analytic methods of ODA allow for perceiving institutions – the cultural attributes of a field – beyond the vague conceptions that are assumed to exist in a field, manifesting via isomorphism somehow in organizational behavior by tying its contents – texts as basic discursive units – to speakers within the field. This is to say that with the organizational field as a starting point, the question of how (and by whom) the laboratory, organizational success, and profession – as key elements of institutional isomorphism – are constructed in the first place can be answered. Thus, my analysis offers theoretical and methodical ways of understanding the cultural attributes – as discourses – that played such a large part in the way that the industrial research laboratory spreads.

The key findings of the discourse analysis supplement the field-based perspective and present the meanings involved in the cultural shift toward science, while also tracing changes to their construction over the seven events from 1895 to 1922. First of all, the discursive perspective illustrates how it is not only field-level action that plays an important part in how concepts, objects, and subject positions are constructed but also their construction is influenced by events and overarching discourses. Field-internal discourses such as disciplinary and professional specialization, the structure of the research landscape, and the development of chemical engineering all impacted the construction of science, chemistry, the laboratory, and its associated actors over the year. Similarly, external events – like WWI – and larger discourses, such as those of conservation, efficiency, preparedness, or the popularization of science, were all used as reference points by discursive speakers, and mostly as points of attachments for discursive constructions: the chemical engineer for conservation, laboratories for efficiency, science for preparedness, and so on. This finding lends support to the hypothesis in translation theory that spreading ideas need not only be changed to fit into new contexts but also that they are changed in ways that render them similar to said contexts (Czarniawska & Joerges 1996; Røvik 2002; Solli et al. 2005). In this way, widespread and legitimized discourses can be used to reinforce new or emerging ones (Phillips et al. 2004, 644), as was done with the bridges built toward the conservation or preparedness discourses.

A second key finding concerns the way that science, research, and chemistry were constructed. Their main attributes can be regarded as superiority over nature and a quest for truth that would lead to societal welfare and material progress – through the industrial, practical applications of science in industry. Science, research, and chemistry were all condensed in the scientific method as the practice of doing science, while chemistry came to be regarded as the fundamental science due to its proximity to the basic building blocks of matter. While the basic attributes that made up these concepts stayed constant over the period analyzed, some features came to prominence when attached to a specific discourse, such as the way that the scientific method could ensure efficient production, or how chemistry was key to preparedness for war. In the early years, the practical applicability of scientific truth was emphasized, while for the later years a simple sequential model was claimed, with pure science creating the necessary building blocks that would be put to practical – industrial – use by applied chemistry and industrial research. Especially in their clear relation of superiority to traditional rule-of-thumb methods as carried out on the workbenches of machine shops and factory floors, the construction of these three concepts shows clear evidence of them being positioned as sources of expertise and the locus of innovation.

Instead of keeping inventors in their workshops on retainer, or hiring individual scientists, chemical corporations established laboratories within their organizations, which becomes understandable once the construction of the laboratory as a discursive object is known. The laboratory's attributes were clear and remained largely fixed over the whole period: It is the place where the scientific method was carried out by university-trained men. Thus, if one wants to harness the proclaimed powers of chemistry, it is not sufficient to simply hire chemists – they need to be placed in a laboratory where they could properly apply the scientific method. Moreover, the laboratory was not conjoined discursively with the university. While it was continuously emphasized that not only research but also teaching was a laboratory-based activity, its whereabouts remained nonspecific, making it possible for industrial organizations to transport the object discursively into their organizations without contradictions that local actors in those corporations had to bridge in their translations. Widespread acceptance is underscored by differentiation into laboratory types, and more and more discussions broaching the issue of proper laboratory setup, equipment, workflow organization, and so on. Understanding what "laboratory" meant to the actors in the field around 1900 adds to the puzzle of the emergence of industrial R&D, especially by highlighting why a laboratory setting was chosen – a decision that may seem obvious today but was not in the realities of those organizations more than one hundred years ago.

Third, the hypothesis of the cultural shift posits a deferral of expertise from inventors and craftsmen to scientists. How were the subject positions of scientists and chemists constructed discursively? Consistent with the attributes of concepts and objects, the scientist and by extension the chemist were regarded as men of science who were trained at universities, and it was only them who could apply the scientific method in research and pass this knowledge on as teachers. They were cast as possessing special personal qualities such as truthfulness and accuracy, and were often rhetorically depicted as pioneers and surveyors – a recurring motif in the American consciousness – of the still unknown expanses of truth. While an air of genius was often attributed to the forefathers of chemistry, its current practitioners were seen instead as methodical practitioners of science. Chemists, and especially the research chemist once the subject position began to differentiate internally, were constructed as greatly superior to foremen, craftsmen, and factory workers whose improvements to products and processes could only be quick fixes, since they were ensnared by tradition and craft knowledge. Discursive attachment to the discourse on science popularization saw the chemist in opposition to managers and executives, who would not value their work, which led to a managerial side being added to the chemists' possible roles in industry. Clearly, the discursive fragments analyzed show how with the scientist, the chemist, and the chemical engineer, the discourse offered subject positions that could be imported into industrial employ, later leading to the creation of the new subject position of industrial chemist.

Lastly, a finding that relates to discourse theory: The analysis showed how concepts, objects, and subjects positions were often related to each other by placing them in specific discursive locations. Thus, the scientific method was materialized in a specific place: the laboratory. The laboratory itself was further set into a specific relationship with the factory characterized by differences of scale, as the lab dealt with the tiny world of molecules, and the factory with the large quantities of materials involved in industrial mass production. Localization also touched upon the motives and outcomes of the places of science, that is, while the practices – the scientific method – were the same, what was pursued within laboratory walls could vary according to the location: university or corporation. This can be regarded not only as relating concepts and objects to each other but also as reactions to events and developments in the field. For example, the heavy emphasis on scale differences between laboratory and factory can also be interpreted as a discursive safeguarding of the laboratory's claim to superiority over nature by way of localizing, in an organizational field where industrial laboratories did not immediately turn out to be houses of magic, as evidenced by the pioneering laboratories.

In summary, the organizational discourse analysis on the meanings assigned to science, the laboratory, the scientist, and related concepts surveys the oft-neglected cultural layer of the organizational field of corporate R&D, and lends credence to the hypothesis of a cultural shift that put science in a position of epistemic authority in the United States of the outgoing nineteenth century. It further illustrates the parts played by fit and appropriateness in a diffusion process, by stressing how the idea of corporate research was translated with a dependence on the organizational environment (as shown in the histories of the pioneers), events, and overarching discourses in efforts to build and reinforce legitimacy of the new idea, as theorized in diffusion theory and Scandinavian Institutionalism (Czarniawska & Joerges 1996; Czarniawska & Sevón 2005). Especially by supplanting the notion of diffusion for translation, the hybrid and constructive elements of diffusion processes are highlighted. By nature of investigating the processes of social construction, the organizational field's institutions are not relegated to hazy, distant rules and norms; instead, their very makeup as definitions of reality that influence organizational decision-making by defining what is perceivable and sayable is shown. In addition, the foundations of the institutional effects of mimetic and normative isomorphism are revealed. Mimesis works as the imitation of practices perceived as successful in order to attain legitimacy, whereas normative isomorphism operates through the development of professional standards that come to be expected. While the discourse analysis did not reveal the construction of any distinct vanguards of the field in terms of specific organizations – not even the R&D pioneers – the way that the scientific method and the laboratory were constructed and came to be expected elements of the organizational structure in the 1910s illustrate how practices such as establishing a laboratory came to be normalized and charged with notions of rationality, efficiency, and superiority, in order to be made fit for widespread diffusion through mimetic isomorphism. The building blocks for normative isomorphism, in turn, can be found in the subject positions of scientists, chemists, and chemical engineers and their positioning vis-à-vis foremen and managers. Lastly, analyzing the emergence of corporate R&D through a lens of diffusion studies expands previous work on the topic by making parts of the cultural layer visible longitudinally and relating it to relevant actors and events, thereby highlighting how the idea of the industrial research laboratory could travel successfully throughout this organizational field.

Science, Laboratories, and Scientists in Sociology and History of Science

Let me now view the results not only through the lens of my study but also relate them to the literature on sociology and history of science, to discover how their insights further frame and enrich the discursive findings, especially considering the importance of place for the laboratory, and the struggles for authority of the chemist and chemical engineer.

In regard of concepts, "science" and "research" acquiring a somewhat magical quality (Daniels 1971, 290ff) and the scientific method reigning supreme in the progressive ideology (ibid., 288ff) certainly fit both the discursive finding and historical writing on the topic. Progress saw itself linked to science as a central part of the "dominant American faith", with material welfare the results of its applications (Rae 1979, 249; LaFollette 1990, 9). Carlson similarly sketched how science promised supremacy over nature in a planned, methodical manner in his explanation for the cultural shift from inventor to scientist (Carlson 1997, 215ff), and Dennis (1987, 495) noted the centrality of the methods of producing scientific knowledge over its contents.

In regard of the central object of interest, the laboratory, it can be said that studies on the emergence and genesis of laboratories underline first and foremost how laboratories were *made* places of science and how associations we have today did not hold true centuries ago. For a twenty-first-century observer, of course, chemistry happens in a laboratory full of experts clad in white, wielding test tubes and Bunsen burners – yet not only the equipment and staff, but also the very placement of science inside the walls of a laboratory are highly contingent and need to be understood as a result of social and cultural effects in which the industrial research laboratory eventually played a part as well.[1] How the experiment came to be placed within a laboratory, the relation of nature to the practices within a laboratory, and the genesis of a distinct place set apart for a laboratory have received particular scrutiny in historical work (cf. Ophir & Shapin 1991; Crosland 2005; Kohler 2008; Jackson 2016). This is illustrated by a line of detailed case studies of laboratories (or scientific spaces) from the seventeenth century onward (cf. James 1989, also Garcia-Belmar 2014), of which only several key aspects that relate to the discourse analysis shall be mentioned here.

Shapin's essay on Robert Boyle's laboratory outlines how belief in the new, unfamiliar methods of experimental science was fostered through attachment to existing institutions such as gentlemanly conventions of access and witnessing, thereby carving out a space for experimentation (Shapin 1988, cf. also Gooday 1998).[2] Gooday (2008) showed how in the nineteenth century laboratories moved out from their parent institutions – museums – as a result of the specialization of science, higher education, and disciplinary ambitions for scientists (ibid., 787), while noting that especially in the nineteenth century laboratories could be hard to distinguish from kitchens, workshops, or museums, further illustrating how the lab was merely one of many possible locations of scientific practice. Crosland (2005) outlines how famous chemists' laboratories, for example, that of Lavoisier, served to strengthen the association of chemistry with the organizational entity of the laboratory, making universities that wanted to "embrace chemistry for teaching and/or research" realize they "needed to provide laboratories" (ibid., 245). By the late nineteenth century, institutional laboratories had become defining features of all scientific disciplines (Jackson 2016, 300). One feature of the modern laboratory that was a result of these struggles for place and credibility is "placelessness". As Kohler (2002) argues, the power of laboratories lies in their separation from nature: "Labs are all the same (more or less) and constitute a universal cultural space; that is why we trust the knowledge produced in them to be universally true" (Kohler 2002, 473). Thus, placelessness "marks the lab as a social form that travels and is easy to adopt, because it seems rooted in no particular cultural soil, but, rather, in universal modernity" (Kohler 2008, 766). We could see the trajectory of the laboratory then as moving from specific spaces that lend credibility to a cultural unit – an organizational form that today has become a non-place, a template, whose powers depend not on genius or method, but cultural practice and extension of the "boundaries" of the laboratory, as argued by Latour (cf. Latour 1983). The construction of the laboratory as a placeless space of science that could be housed at universities or corporations is also commented upon by Dennis, who remarks that the key accomplishments of entrepreneurs such as Steinmetz (GE) and Jewett (AT&T) were the realization that the laboratory, the heart of the research school, was "portable" (Dennis 1987, 505).

These briefly sketched concepts serve to illustrate the importance of place in the development of the laboratory, a category that can be found twofold in the way that the laboratory is made into a discursive object: on the one hand, as the physical manifestation of the scientific method and experimental practice, and on the other in placing the laboratory in relation to the factory as a matter of scale. The early contributions to the discourse, establishing the laboratory's attributes as outlined, certainly fit the category of placelessness, locating the laboratory neither at the university nor in the corporation, but instead as an abstract space of experimentation where nature could be reproduced and subjugated – in this way, the laboratory became the materialization of a concept with all that it entails. Yet, as Kohler perceptively notes, "placelessness, like all cultural conventions, runs in cycles and is dependent on prevailing – but fickle – cultural weather" (Kohler 2002, 495). Perhaps the questions of laboratory scale can be understood in this way, as analogous to the troubles of morphology he described. When the ability to reproduce nature in the industrial laboratory was questioned (not only discursively but also by executives demanding results in the form of products and processes, which the early laboratories could not produce as fast as those believing discursive claims may have believed), its placelessness – that is, the cultural conventions governing the credibility of science – was under attack. Positioning the industrial research laboratory in relation to the factory, from small to large, can perhaps not be seen as re-naturalizing, since it did not necessarily change the laboratory's relation to nature, but as repositioning the laboratory within the industrial organization. In this way, its claims of control and supremacy were safeguarded, but by adding distance – literal distance, since early laboratories made it a point to be exempt from routine and shop work, as well as conceptual distance, in the need to scale the small laboratory world to the large factory, thereby opening up the cultural practice of laboratory research to the entrance of the chemical engineer, and transforming subject positions in the process. Furthermore, the key examples from laboratory history cited help us in understanding the laboratory in its physical setting, its apparatus and equipment, as well as the reasons put forward for establishing laboratories as a cultural product. With regard to the "whys" of laboratory establishment probed earlier, an interesting hypothesis is proposed by Kohler:

> An analogous argument may explain why labs and lab science came to have such a prominent place in modern industrial corporations: the analytic categories and practices of lab science were congruent with the new managerial hierarchies and procedures of large-scale industrial capitalism, whereas those of the older shop culture were not. As well, scientists proved useful allies for modernizers striving to transform the traditional business firm into the modern managerial leviathan.
>
> (Kohler 2008, 768)

The rationales outlined previously – utility, application, profit, and a certain security against the forces of the market by pioneering efforts – as well as the construction of the laboratory as filled with scientifically trained experts certainly fit the analytic categories of corporate control introduced by Fligstein and others, including the historically enriched case studies on the pioneers.

Lastly, what about the men of science? At the end of the nineteenth century, the pace of professionalization was ever increasing, with professional societies proliferating and patrolling the boundaries of their respective disciplines (Mody 2016, 164), and the vocation of science having become a career by then (White 2016, 154). But the "usefulness of knowledge, and the relationship between science, commerce, and industry, were vexed and controversial" (ibid., 155), especially with old ideas of genius and the powers and proper role of science challenged and transformed with the beginning of the Progressive Era (cf. Daniels 1971, 288ff). Thus, while we can certainly see an achievement of professional autonomy in academic chemistry by the turn of the century reflected in institutional and organizational arrangements (Daniels 1976, 77f), it is not that simple in the sphere of ideas and concepts.

Understanding the entry of more and more chemists into the industrial workforce purely as a linear extension of academic chemistry's plight in a different domain would fall short of the empirical reality observed in the discursive texts earlier. Instead, we could see how scientists needed to legitimize and safeguard their expertise from rival groups using not scientific but craft knowledge (cf. Mody 2016, 167), or the mysterious practices involved in alchemy. In keeping with the imagery of localization and places of science carved out previously, a useful way of understanding such conflicts and, subsequently, why a subject position was constructed in a certain way, is to look at struggles for authority and the lines of demarcations drawn within them. The concept of boundary-work with its special focus on the "maps" drawn of (and around) science helps in assessing the discursive reality described earlier. In introducing the concept of boundary-work, Thomas Gieryn is not interested in some abstract, universal quality of science from which it receives its epistemic authority,[3] but rather how the boundaries are drawn around and within science account for the legitimacy of science-based knowledge in modern societies (Gieryn 1999). Instead of looking at the specific contents of science, Gieryn proposes analyzing science's representations that occur "downstream" (Gieryn 1999, 27), that is, every time the authority of science is called into question.

Boundary-work, then, denotes the specific practices of creating, maintaining, and repositioning those demarcations drawn around science, which mainly happens through discursive activity. In doing boundary-work, actors connect to previously established "maps" of science (here: other discourses) and utilize this repertoire of familiar characteristics for further legitimacy (cf. Gieryn 1983, 783; 1995, 407). Gieryn distinguishes three types of boundary-work: expulsion, protection of autonomy, and expansion.

Episodes in the construction of the chemist's subject position can be regarded as a case of the expansion of boundaries, with chemists claiming to know better than the alchemists of old (Nummedal 2016, 59), or the late nineteenth century's foremen and inventors. In this way, the characteristics of the concept of science needed to be attached to the subject position to illustrate its epistemic authority: domination over nature, the use of the scientific method, and especially higher education and expertise. Gieryn (1983, 784ff) shows how differing boundaries were constructed toward the various groups. The discursive reasoning involved with the different groups that the industrial chemists found themselves faced with can be understood along similar lines. Thus, scientific instruction yielding insight into fundamental principles was emphasized vis-à-vis those hostile foremen and plant managers, who were "mere" practical men. On the one hand, only whoever mastered the theoretical architecture of nature could hope for improvements in products and efficiency. On the other hand, managers seemed leery of "impractical" chemists, far from applications and any feeling for costs and profits – which is why the boundary was drawn differently in this case. Here, the need to "sell" the value of research was recognized, to translate it into the language of business, while claims to theoretical knowledge were downplayed.

The notion of boundary-work also furthers our understanding of the chemical engineer's subject position. Chemical engineers expanded their domain over territory previously claimed not only by engineers but also by industrial chemists and analysts, sometimes emphasizing their chemical training (versus engineers), or their engineering and managerial abilities (versus industrial chemists and analysts), while drawing upon other, established discourses such as conservation or efficiency. In the genesis of chemical engineering organizations, one could argue that another type of boundary-work becomes visible, which – instead of delegitimizing another's claims to scientificness – stresses how both are sciences with their respective places in the field, as was needed in the case of chemical engineering with regard to chemistry, to not act hostile against the larger and more powerful ACS.

Thus, the notion of boundary-work further elucidates the findings for several reasons. First, it highlights how there is and was never any fixed idea of science that was transformed (or even perverted) with the advent of industrial laboratories; the discourse analyzed earlier serves instead as an instance of one of science's numerous struggles for demarcations, drawing boundaries differently from case to case. Boundary-work also highlights agency in the role of different kinds of rhetoric and discursive contribution involved in constructing concepts and especially subject positions, as in legitimate members of science. Lastly, by pointing toward different repertoires used to delineate different boundaries, Gieryn helps to classify the (industrial) chemist's and chemical engineer's constructive struggles in their different environments facing different oppositions, going far beyond the mere presupposed notions of pure scientists at university and applied scientists in industry. With our interests and research question in mind, the empirical – discursive – material could be probed for constructions and transformations, for boundaries drawn and re-drawn, without being guided by the preset notion of purity and applications.

Discussion

Implications for the Diffusion of Innovations, Sociology of Organizations, and History of Science

This study used a combination of theory from the diffusion of innovation and organizational sociology, methodical insights from discourse theory, as well as historical data that informed key junctures in the ways these diverse parts were put together, yielding new insights on the genesis of corporate R&D. But the results of scientific research should not simply be used to enhance knowledge about a specific subject. Instead, going back to theory and method, results gathered here shall be used to assess the interplay of theoretical components. What the research design and findings mean for their respective approaches will be discussed in the following.

Recontextualizing Diffusion

What does the union of methods and theory employed here add to studies of diffusion and translation? In this brief discussion, I want to focus on theoretical contributions made by combining diffusion with institutional and translation theories, and on the role of history in diffusion analysis.

The history of the industrial research laboratory further underscores the need to analytically incorporate the environment or social system that a given innovation is diffusing in. Framing the research laboratory as an innovation offering some perceived advantage to potential adopters could not have explained its spread and subsequent institutionalization, especially since – looking at the history of the early laboratories – breakthroughs were lacking as the labs were falling far behind the claims of their founders. Instead, by shifting focus toward the cultural attributes of the organizational field, my analysis shows how the perceived advantage of laboratories and scientists was institutionally enacted in a collective process of discourse construction. Supplementing diffusion studies with some form of *cultural* analysis – which is not limited to organizational institutionalism, of course – will also aid in probing the thin line between innovations spreading, and innovations staying, or understanding why some remain fads, while others become standards. As Colyvas and Jonsson (2011, 29, Table 1) outline, the intersection of diffusion and institutionalization may lead to innovations that diffuse and become ubiquitous but are not accepted institutionally, contrasted with those that diffuse and become institutionalized resulting in widespread acceptance and appropriateness. The authors stress that legitimacy is not sufficient for institutionalization; instead, they theorize that the modes of reproduction and the diversity of links of a spreading innovation are of prime importance (ibid., 43ff). My analysis shows how through the collective, organizational "voice" of chemists and scientists in general, the discourse on science and laboratories could be reproduced time and time again, leading to a normalization of the practice of establishing laboratories. Furthermore, the diversity of links – as embeddedness with "higher- and lower-order frames, rules, and routines" (ibid., 44; see also Alasuutari 2015) – was achieved through various translations of the spreading practice.

Regarding the notion of translation, the case of organized R&D in the American chemical industries illustrates why it is rarely if ever fixed entities that spread in a diffusion process. Instead, ideas, practices, and things change while moving through a field in an active process of editing (Sahlin & Wedlin 2008; see also Creed et al. 2002). This not only allows for bringing the actor or groups of actors back in, beyond groups of adopters such as innovators, early adopters, and laggards but also enables the analysis to include the positions that actors inhabit, and their respective interests and *Lebenswelten* by staying alert to different motifs, goals, and social realities that various actors or groups may have that account for translating an idea in a certain way. In addition, the spread of industrial research laboratories also shows how translation acts on different levels: the local and the global. For the local, it is the different way the laboratory was carried into the corporation, different laboratory setups, places within the organizational structure, different staffing strategies, and different research policies exhibited by the research pioneers' labs. From a macro-perspective, it can be seen that the entity "R&D laboratory" diffused, but only detailed case histories inform the varied ways that it had to be translated in order to fit. In the many associations of the laboratory (and other relevant concepts and subject positions) with larger discourses and other events, the practice of establishing R&D laboratories was translated on a global scale, changing its scope, mission, form, raison d'être, and so on.

These findings also add to the theoretical ensemble put forward by Scandinavian Institutionalism. Scholarship utilizing the translation metaphor often proposes some form of shared, higher-order cultural resources that set rules for editing and that local translations have to abide by, in order to make the further spread of a specific translation possible (Sahlin & Wedlin 2008). These resources are generally called "master ideas", as they are higher order cultural accounts that provide collectively shared resources for legitimation (Czarniawska & Joerges 1996, 36; see also Meyer 1996). Although often assumed in studies understanding diffusion as translation (Berglund & Werr 2000; Näsie & Rohde 2007), what these master ideas are and how they are used remains sketchy at best. Using the organizational field as a boundary-setter and discourses as the content of a field's culture, my analysis supplies clear means to get these master ideas in focus, while also providing hints as to how they are translated in practice, as the various ways that other discourses were utilized in the construction of concepts, objects, and subject positions have shown. As can easily be seen, diffusion theorizing is greatly enriched by substituting diffusion for translation, and by combining its insights on the relevance of cultural variables to the institutional theory that further elucidates what said culture is made up of, and how it is enacted.

Lastly, what about the role of history in diffusion studies? As Djelic (2008) outlines, there are several approaches within diffusion studies toward historical data. These data can either be used as objectified and measurable "chunks" that form a succession of events, these "stones and bricks" (ibid., 548) constituting the building blocks with which preexisting theoretical frames are tested. Such an approach emanates from a very positivistic understanding of history. Next, starting from an understanding of historical data as thick, continuous, made up of complex narratives, and informed by interpretive approaches such as the Annales school (Burke 2015), history is used to generate theoretical propositions (Djelic 2008, 549). But essentially, Djelic argues that historical data needs to be used in a third way in diffusion studies, where translation and the importance of culture is taken seriously by making acts of constructions visible: "[W]e cannot understand diffusion without understanding historical construction" (ibid., 551). Only by being aware of the constructive nature of language and the necessary plurality of narratives involved at any historical juncture can diffusion studies properly utilize historical data. Historians' interpretations of events and trajectories can then be used as starting points to access primary data – the actions and voices of actors at the time – that forms the basis for making conflicting narratives and with it the construction of social reality itself visible. Due to the increasing digitalization of large amounts of historical material, the methodical arsenal will have to be developed much further to cope with the amounts and types of these data, especially to be able to exceed highly detailed historical reconstructions by making more general mechanisms visible, such as in this case, where the constructive functions of discourse impacted the diffusion of an idea in various ways.

Organizations as Texts

Organizational discourse has become more and more relevant to the study of organizations in the last two decades, partly due to a "growing disillusionment" (Grant et al. 2004, 1) with conventional theorizing, and partly due to the incorporation of the linguistic turn in the social sciences and organization studies (Deetz 2003). A growing number of publications and special issues of journals (Grant et al. 2004, 2005) are testament to this trend. Of course, as the definitional struggles in theory and method discussed previously revealed, neither clear definitions nor common approaches exist in ODA so far. Organizational discourse studies are loosely connected by the general idea that there is such a thing as discourse – made up of texts – that composes organizations, while these texts are in turn created by organizations (Hardy 2001, 26). Coming from this base definition, two approaches can be distinguished: analyses of discourse *within* organizations, such as memos and plans (Hardy 2001, 34); and analyses of discourse that surrounds organizations and shapes not only their possibilities of action but also their very form and existence itself, such as the discourses on business re-engineering or total quality management (Hardy et al. 2005; Green et al. 2009). The analysis undertaken here falls firmly within the second group, even though organizational text-production was in focus, the functioning of discourse uncovered how relations between texts, events, and other discourses (intertextuality and interdiscursivity) shape the space of what is sayable and doable, as manifested in discursively constructed institutions.

I believe the way that my analysis relates to and enhances the current work in ODA is twofold, based on the fundamental assumptions of discursive research programs. Alvesson and Kärreman (2000b, 1135, Figure 2) describe the area of conflict in which studies of organizational discourse reside: casting discourse as either determining, or having only a softer effect on social reality, and focusing on the local-situational context versus zooming out to capture the macro-context through grand discourses. Locating a study in either of the resulting quadrants results in theoretical and methodical consequences, such as the definition of discourse and the nature of the material that is to be analyzed. My analysis brings the macro-context of corporate innovation into focus via the notion of the organizational field, while utilizing a strong notion of discourse that is assumed to construct the contents of social reality, such as organizations and actors in the first place. The theoretical and methodical consequences of this research design can, of course, be challenged, but they also aid in laying the groundwork and clarifying terms.

For one, too often the tenet "discourse constructs" is taken for granted, instead of being the starting point of a sensible analysis of discourse (Alvesson & Kärreman 2011a, 1199; Iedema 2011, 1168f). How exactly reality is constituted through language, and how it gains massivity in institutional practice, should form the basis of this kind of research program. At the heart of such concerns lies the question of discursive agency (Hardy 2004; Cooren 2004; Iedema 2011): How can collections of texts *do* things, and what are the limits to language? Similar to ideas not traveling on their own, texts do not write themselves – so some form of textual agency is supplied through the connection to discursive speakers who inhabit certain positions in the field that allow them to spread texts far and wide and enforce their contents (Phillips et al. 2004). My analysis has highlighted what form of reality – in the manifestation of concepts, objects, and subject positions – was described and thus constructed by speakers, which connected to the subsequent spread of organized R&D as the consequences of these constructive effects. Of course, the actual effects that these linguistic constructions had could only be attributed after the fact via indicators such as organizational development in academia and industry, the proliferation of the laboratory, and the individual shapes the laboratories took. By combining an ODA approach with institutional theory, the case for discourse constructing reality is strengthened by understanding institutions and their effects, such as isomorphism, as being made up of and shaped by discourses. This combination is a powerful one and could be harnessed further by zooming in on the precise relationship between discursive and institutional effects, solidifying our understanding of how discursive constructions of reality are enacted in institutionalized practices, and going further and into more detail than other approaches that combine institutions and discourse, such as discursive institutionalism (Schmidt 2008).

In addition, regarding the dimensions of micro- and macro-context, what the context of discourse is, how it influences discourse, and where it starts and ends is often left vague in ODA. Keenoy and Oswick lament how by drawing too strict of boundaries around context and extracting discourse from it, a partialized and de-contextualized idea of discourse is reified:

> There is a tendency within discourse analysis to place parameters (that is, create a contained space) around a piece of discourse. *In extremis*, this process involves a robust delineation of text and context where the focal discourse is uncoupled and investigated independently of the physical surroundings and the wider social context in which it occurs.
>
> (Keenoy & Oswick 2004, 139, emphasis in original)

In their view, context is flattened to the backdrop instead of being depicted as being in a mutual relationship with discourse, as both shape discursive texts while also being embedded in the discursive episode itself (Grant et al. 2004, 22). Although Keenoy and Oswick propose the somewhat diffuse notion of "textscape" (Keenoy & Oswick 2004, 140f) to deal with said embeddedness, the concepts borrowed from organizational institutionalism, namely the organizational field, serve to gain access to discursive units without drawing too strict of boundaries or negating their embeddedness. This is due to the hybrid nature of the field as a *social* space of perception and expectations, instead of being bordered by time and physical location. Moreover, the three-dimensional approach relates discursive texts to their context of production and reception, which are elucidated through the field, by pointing out the actors' positions and their relations to each other that influence visibility and the "weight" of discursive contributions. In summary, my study extends the notions of organizations and their entities as made up of text and talk (cf. Brunsson 1997) while refining theoretical and methodical notions through the aid of organizational institutionalism – a hybrid approach that opens up many more avenues for theory building and method development.

Social Science

The detailed analysis of the organizational field of corporate R&D in the chemical industries, and its discursive currents, serves as another puzzle piece in the history of one of the most important novelties for both science and industry that shaped science as a vocation, as a profession, and in its disciplinary identities until the present day. In this way, my analysis adds insights to both business history and the history of science. Much of the history of industrial research was viewed through a lens focusing on what science *is* and how its identity manifests in who its practitioners are and what they do, often being tied to the distinction of pure and applied science or later on basic and applied research (Godin & Schauz 2016; Kaldewey & Schauz 2018). The opposition of pure and applied as cognitively different types of research was often coupled with the assumption that they correlated with institutional settings – university and industry. These ideas formed what might be called a dominant discourse in the social sciences, only filtering phenomena through this slant. Coming from an ideology of the purity of science, Mertonian norms, or the presumption that science as a system and a vocation was inherently special, of course, researchers would regard industrial science as a challenge to the nature of science and the scientist. Shapin summarizes these problems aptly (Shapin 2008, 114ff) and remarks how accounts from outside these laboratories were often "celebrations, denigrations, or expressions of various anxieties" (ibid., 99). Hence the significant focus of cultural attention that industrial science received by the 1920s (ibid., 95), which is also mirrored in the discursive texts that show how industrial research and the industrial laboratory came to be established constructs toward the end of the period analyzed. As the analysis shows, to actors of the time the distinction of types of science played a minor role in the discourse on science, gaining steam only toward the 1920s when the laboratory was institutionalized. That is why basing the analysis on the "dominant discourse" was rejected here, in order to develop a more nuanced view that highlighted constructions instead of preconceiving them.

Two key insights can be taken from these thoughts in the context of the sociology and history of science, one about how we analyze scientific pursuits, and the other about the narratives constructed about them. First, the analysis illustrates how what Gieryn calls the "downstream approach" can be a powerful tool to understand the inner, social workings of science: "Epistemic authority is decided downstream from all that, as claims float through layers of cartographic interpretations where credibility is attached or removed" (Gieryn 1999, 27). Starting at a similar vantage point as the cultural shift toward science, he questions how science came to be the source of authority it is today. He argues that to find sociological answers, instead of looking into what scientists do, their representations of science brought up once science's credibility is questioned, "downstream" of the actual practices of science, need to be analyzed (ibid., 25). The early years of the industrial research laboratory form such an episode – or many episodes – of boundary-work, clearly exhibiting how "science" was instantiated discursively and shaped by local and occasional circumstances, be they topical discourses sweeping through the field, or local opposition within individual corporations. Players, stakes, and their interests are mapped with a theoretical framing added that allows an understanding of the boundary-work beyond individual actors and their argumentative contributions, as well as how organizational and institutional arenas shape the forms and contents that claims about science take. Of course, what science is, why it happens in a laboratory, and what a scientist can do are not the only topics of boundary-work of interest here. There are other demarcations drawn in this episode, such as those regarding the individual scientist and the group, or the discursive conflict over scientific discovery versus invention that resonated strongly in patenting practices and the plight of independent inventors (Miller 2011). In addition, shifting from the contents of truth-claims about science toward players and arenas would yield further insights, especially considering industries, as Smith (1990, 122) remarks, whose business does not closely align with scientific disciplines. How would the discursive trajectories have looked with many more players vying for epistemic primacy and a much more complex cartography of the sciences? In this way, understanding science's epistemic authority and the success of the industrial laboratory through the way it is contested discursively offers more fruitful insights than intense scrutiny of laboratory practices as suggested by approaches informed by Science and Technology Studies.

My second point is a more nuanced one and concerns the stories we tell about science. Analyzing independent invention, Whalley argues that its decline is usually embedded within two stories: the rise of the large, integrated corporation, as told by economists, and the emergence of science-led innovation, as related by historians of science and technology. In a nutshell, he argues that the inventor was displaced twice: from the market for innovation back in the late 1800s, and from academic interest through these convincing "meta-narratives"

(Whalley 1991, 210; see also Dennis 1987). A related point is made by Shapin:

> That the emergence of organized industrial science so intruded itself into American consciousness arises partly from the novelty of the phenomenon and partly from the challenge it represented to existing understandings about the nature of science and of the scientist.
> (Shapin 2008, 99)

Thus, any analysis of such a topic needs to be aware of the viewpoints taken, which does not just apply to the historical material, such as the discursive texts analyzed here. The topics and explanations put forward by academic scholarship itself – topics denoting the R&D lab as a revolution, an innovation, or a pivotal shift – are also exposed to definitional struggles, boundary-work, and explanatory fashions, that need to be considered in order not to fall prey to pro-innovation biases and only write winners' histories – winners that, in retrospect, often seem inevitable. The whole research design of this study started with an interest in language and diffusion that was subsequently informed by a hunch about the cultural shift of epistemic authority toward the scientist – the key "meta-narrative" followed here, even though others (economic, legal, and organizational) were included for the multi-perspectival construction of the organizational field. In the end, one needs to be aware of the narratives one partakes in (or subscribes to) when trying to further elucidate such an episode of innovation, be it framed as successful diffusion, organizational change, or boundary-work. When using a social constructivist ontology that is aware of the creative effects of language, taking a reflexive stance is key.

In the end, only further analysis can highlight the ways that discourse – as a force both creating and legitimizing reality – influences organizational reality in its respective fields. Furthermore, the analysis shows the merit of inspecting discursive dynamics on a grander, yet more grounded, level, without resorting to hinging powerful claims on only few paradigmatic texts, and without assuming any identities (such as the "pure scientist") and their respective interrelations beforehand, instead showing how the paradigmatic texts are made widely heard, and how the said identities are constructed in the first place. In this way, the pitfall of assuming every meaning to be discursive is evaded by explicitly scrutinizing *how* discourses construct meaning (Alvesson & Kärreman 2011a, 1199). Extending the analysis toward WWII and beyond could further probe the translations undergone by the industrial research laboratory while it became the de-facto standard of corporate innovation, leading to a better understanding of how exactly isomorphic pressures manifest textually and are in turn transported into organizations. There are still many ways in which the organizational research laboratory can serve to increase our knowledge of history, the poignancy of our theory, and the precision of our methods – or, as Gieryn (1999, 35) put it: "Lots of work ahead".

Notes

1 A striking illustration of the contingency of the association of science with a laboratory setting can be found in Secord (1994), who shows how existing class structure in England of the nineteenth century placed artisan botany in a rather surprising place – the pub.

2 It needs to be noted that the relationship of the laboratory to the experiment is highly complex and can only be outlined cursorily here (cf. Knorr-Cetina 1992; Weingart 2003, 67ff). Klein contests Shapin's clear equation of experimental practice with a laboratory, emphasizing instead how early laboratories, although distinctly chemical in nature, were also artisanal workplaces (Klein 2008); or, as Jackson notes, the "laboratory was not, in the first instance, a place of experiment. It was the particular space in which the technical hazards of chemical practice, whatever its purpose and setting, could be made to conform to social norms" (Jackson 2016, 299).

3 "Epistemic authority" here means "the legitimate power to define, describe, and explain bounded domains of reality" (Gieryn 1999, 1), one of the central characteristics of modern science.

References

Ackerman, Carl W. 1930. *George Eastman*. Boston/New York: Houghton Mifflin Company.

Aftalion, Fred. 1991. "A History of the International Chemical Industry." In *The Chemical Sciences in Society Series*, edited by Arnold Thackray. Philadelphia: University of Pennsylvania Press.

Alasuutari, Pertti. 2015. "The Discursive Side of New Institutionalism." *Cultural Sociology* 9 (2):162–184.

Alvesson, Mats, and Dan Kärreman. 2000a. "Taking the Linguistic Turn in Organizational Research." *The Journal of Applied Behavioral Science* 36 (2):136–158.

Alvesson, Mats, and Dan Kärreman. 2000b. "Varieties of Discourse: On the Study of Organizations Through Discourse Analysis." *Human Relations* 53 (9):1125–1149.

Alvesson, Mats, and Dan Kärreman. 2011a. "Organizational Discourse Analysis - Well Done or Too Rare? A Reply to Our Critics." *Human Relations* 64 (9):1193–1202.

Alvesson, Mats, and Dan Kärreman. 2011b. "Decolonizing Discourse: Critical Reflections on Organizational Discourse Analysis." *Human Relations* 64 (9):1121–1146.

American Chemical Society, ed. 1951. *Chemistry... key to better living. Diamond Jubilee Volume*. Washington, DC: American Chemical Society.

American Cyanamid Company. 1937. *After 30 Years. 1907-1937*. New York: American Cyanamid Company.

Angell, James Rowland. 1919. "The Development of Research in the United States." Proceedings of the Thirty-third Annual Convention of the Association of American Agricultural Colleges and Experiment Stations, Chicago, Illinois.

Baird, Spencer F., ed. 1872. *Annual Record of Science and Industry for 1871*. New York: Harper & Brothers.

Baird, Spencer F., ed. 1873. *Annual Record of Science and Industry for 1872*. New York: Harper & Brothers.

Baird, Spencer F., ed. 1874. *Annual Record of Science and Industry for 1873*. New York: Harper & Brothers.

Baird, Spencer F., ed. 1875. *Annual Record of Science and Industry for 1874*. New York: Harper & Brothers.

Baird, Spencer F., ed. 1876. *Annual Record of Science and Industry for 1875*. New York: Harper & Brothers.

Baird, Spencer F., ed. 1877. *Annual Record of Science and Industry for 1876*. New York: Harper & Brothers.

Baird, Spencer F., ed. 1878. *Annual Record of Science and Industry for 1877*. New York: Harper & Brothers.

Baird, Spencer F., ed. 1879. *Annual Record of Science and Industry for 1878*. New York: Harper & Brothers.

Baldwin, Melinda. 2015. *Making Nature. The History of a Scientific Journal*. Chicago/London: The University of Chicago Press.

Barley, Stephen R., and Gideon Kunda. 1992. "Design and Devotion: Surges of Rational and Normative Ideologies of Control in Managerial Discourse." *Administrative Science Quarterly* 37 (3):363–399.

Barley, Stephen R., and Pamela S. Tolbert. 1997. "Institutionalization and Structuration: Studying the Links between Action and Institution." *Organization Studies* 18 (1):93–117.

Bartlett, Howard R. 1941. "The Development of Industrial Research in the United States." *Journal of the Patent and Trademark Office Society* 23:475–513.

Bates, Ralph S. 1945. *Scientific Societies in the United States*. New York: John Wiley & Sons.

Bazerman, Charles. 1988. *Shaping Written Knowledge. The Genre and Activity of the Experimental Article in Science*. Madison, WI: The University of Wisconsin Press.

Beardsley, Edward H. 1964. *The Rise of the American Chemistry Profession, 1850-1900, University of Florida Monographs: Social Sciences*. Gainesville, FL: University of Florida Press.

Becker, Howard S. 2014. *What about Mozart? What about Murder? Reasoning from Cases*. Chicago/London: The University of Chicago Press.

Bensaude-Vincent, Bernadette. 1997. "In the Name of Science." In *Science in the Twentieth Century*, edited by John Krige and Dominique Pestre, 319–338. Amsterdam: Harwood Academic Publishers.

Berger, Peter L., and Thomas Luckmann. 1967. *The Social Construction of Reality*. New York: Anchor.

Berglund, Johan, and Andreas Werr. 2000. "The Invincible Character of Management Consulting Rhetoric: How One Blends Incommensurates while Keeping them Apart." *Organization* 7 (4):633–655.

Birr, Kendall. 1957. *Pioneering in Industrial Research. The Story of the General Electric Research Laboratory*. Washington, DC: Public Affairs Press.

Birr, Kendall. 1966. "Science in American Industry." In *Science and Society in the United States*, edited by David D. Van Tassel and Michael G. Hall. Homewood, IL: The Dorsey Press.

Birr, Kendall. 1979. "Industrial Research Laboratories." In *The Sciences in the American Context: New Perspectives*, edited by Nathan Reingold, 193–207. Washington, DC: Smithsonian Institution Press.

Bogert, Marston Taylor. 1908. "American Chemical Societies." *Journal of the American Chemical Society* 30 (2):163–182.

Bohning, James J. 1989. "The 1893 World's Congress of Chemists. A Center of Crystallization in a Molecular Mélange." *Bulletin for the History of Chemistry* 3:16–21.

Bohning, James J. 1990. "The Continental Chemical Society." *Bulletin for the History of Chemistry* 6:15–21.

Bohning, James J. 2001. "Opposition to the Formation of the American Chemical Society." *Bulletin for the History of Chemistry* 26 (2):92–103.

Bolton, Henry Carrington. 1902. *Chemical Societies of the Nineteenth Century.* Washington, DC: Smithsonian Institution.

Bolton, Henry Carrington (1899). 1967a. *A Select Bibliography of Chemistry 1492-1897. First Supplement.* New York: Kraus Reprint Company.

Bolton, Henry Carrington (1903). 1967b. *A Select Bibliography of Chemistry 1492-1904. Second Supplement.* New York: Kraus Reprint Company.

Bolton, Henry Carrington (1893). 2005. *A Select Bibliography of Chemistry 1492-1892.* Mansfield Centre, CT: Martino Publishing.

Boxenbaum, Eva, and Stefan Jonsson. 2008. "Isomorphism, Diffusion and Decoupling." In *The SAGE Handbook of Organizational Institutionalism*, edited by Royston Greenwood, Christine Oliver, Kerstin Sahlin, and Roy Suddaby. London: SAGE.

Bright, Arthur. 1949. *The Electric Lamp Industry: Technological Change and Economic Development from 1800 to 1947.* New York: The MacMillan Company.

Brinton, Mary C., and Victor Nee, eds. 1998. *The New Institutionalism in Sociology.* New York: Russel Sage Foundation.

Broderick, John T. 1945. *Willis Rodney Whitney. Pioneer of Industrial Research.* Albany, NY: Fort Orange Press, Inc. Publishers.

Browne, Charles Albert, and Mary Elvira Weeks. 1952. *A History of the American Chemical Society.* Washington, DC: American Chemical Society.

Bruce, Robert V. 1972. "A Statistical Profile of American Scientists, 1846-1876." In *Nineteenth-Century American Science. A Reappraisal*, edited by George H. Daniels. Evanston: Northwestern University Press.

Bruce, Robert V. 1987. *The Launching of Modern American Science 1846-1876.* New York: Alfred A. Knopf.

Brunsson, Nils. 1997. "The Standardization of Organizational Forms as a Cropping-Up Process." *Scandinavian Journal of Management* 13 (3):307–320.

Bugos, Glenn E. 1989. "Managing Cooperative Research and Borderland Science in the National Research Council, 1922-1942." *Historical Studies in the Physical and Biological Sciences* 20 (1):1–32.

Burke, Peter. 2015. *The French Historical Revolution. The Annales School 1929-2014.* Cambridge, UK: Polity Press.

Burman, Erica. 2003. "Discourse Analysis Means Analysing Discourse: Some Comments on Antaki, Billig, Edwards and Potter 'Discourse Analysis Means Doing Analysis: A Critique of Six Analytic Shortcomings'." Accessed 2018/10/25. https://extra.shu.ac.uk/daol/articles/open/2003/003/burman2003003-paper.html

Callon, Michel. 1980. "Struggles to Define What Is Problematic and What Is Not: The Socio-Logic of Translation." In *The Social Process of Scientific Investigation: Sociology of the Sciences*, edited by Karin Knorr, Roger G. Krohn, and Richard Whitley, 197–219. Dordrecht: Reidel.

Callon, Michel. 1986. "The Social Process of Scientific Investigation: Sociology of the Sciences." In *Power, Action and Belief: A New Sociology of Knowledge*, edited by John Law, 196–233. London: Routledge & Kegan Paul.

Campbell, John L. 1998. "Institutional Analysis and the Role of Ideas in Political Economy." *Theory and Society* 27:377–409.

Carlson, W. Bernard. 1991. *Innovation as a Social Process. Elihu Thomson and the Rise of General Electric, 1870-1900.* Cambridge: Cambridge University Press.

Carlson, W. Bernard. 1995. "The Coordination of Business Organization and Technological Innovation within the Firm: A Case Study of the Thomson-Houston Electric Company in the 1880s." In *Coordination and Information. Historical Perspectives on the Organization of Enterprise*, edited by Naomi Lamoreaux and Daniel M.G. Raff, 55–99. Chicago/London: University of Chicago Press.

Carlson, W. Bernard. 1997. "Innovation and the Modern Corporation: From Heroic Invention to Industrial Science." In *Science in the Twentieth Century*, edited by John Krige and Dominique Pestre, 203–226. Amsterdam: Harwood Academic Publishers.

Carlson, W. Bernard. 2007. "Toward a Non-Linear History of R&D: Examples from American Industry, 1870-1970." In *The Creative Enterprise: Strategy*, edited by Tony Davila, Marc J. Epstein, , and Robert Shelton. Westport, CT: Praeger.

Carlson, W. Bernard, and Stuart K. Sammis. 2009. "Revolution or Evolution?: The Role of Knowledge and Organization in the Establishment and Growth of R&D at Corning." *Management & Organizational History* 4 (1):37–65.

Carmichael, Emmett B. 1974. "Fifty Golden Years: History and Accomplishments of the American Institute of Chemists." *The Chemist* 51:22–23.

Cerveaux, Augustin. 2013. "Taming the Microworld: DuPont and the Interwar Rise of Fundamental Industrial Research." *Technology and Culture* 54 (2):262–288.

Chalaby, Jean K. 1996. "Beyond the Prison-House of Language: Discourse as a Sociological Concept." *British Journal of Sociology* 47 (4):684–698.

Chandler Jr., Alfred D. 1959. "The Beginnings of "Big Business" in American Industry." *Business History Review* 33 (1):1–31.

Chandler Jr., Alfred D. 1965. "The Railroads: Pioneers in Modern Corporate Management." *Business History Review* 39 (1):16–40.

Chandler Jr., Alfred D. 1977. *The Visible Hand. The Managerial Revolution in American Business*. Cambridge, MA/London, England: Cambridge University Press.

Chandler Jr., Alfred D. 1990a. *Scale and Scope. The Dynamics of Industrial Capitalism*. Cambridge, MA/London, England: The Belknap Press of Harvard University Press.

Chandler Jr., Alfred D. 1990b. *Strategy and Structure: Chapters in the History of the American Industrial Enterprise*. Cambridge, MA/London, England: MIT Press.

Chandler Jr., Alfred D. 2005. *Shaping the Industrial Century. The Remarkable Story of the Evolution of the Modern Chemical and Pharmaceutical Industries*. Cambridge, MA/London, England: Harvard University Press.

Chandler Jr., Alfred D., and Stephen Salsbury. 1971. *Pierre S. Du Pont and the Making of the Modern Corporation*. New York: Harper & Row.

Cochrane, Rexmond C. 1978. *The National Academy of Sciences. The First Hundred Years 1863-1963*. Washington, DC: National Academy of Sciences.

Coe, Brian. 1976. *The Birth of Photography. The Story of the Formative Years 1800-1900*. Cambridgeshire: Ash & Grant.

Cohen, Arthur M., and Carrie B. Kisker. 2010. *The Shaping of American Higher Education. Emergence and Growth of the Contemporary System*, 2nd ed. San Francisco, CA: Jossey-Bass.

Collins, Douglas. 1990. *The Story of Kodak*. New York: Harry N. Abrams, Inc.

Colyvas, Jeanette A., and Stefan Jonsson. 2011. "Ubiquity and Legitimacy: Disentangling Diffusion and Institutionalization." *Sociological Theory* 29 (1):27–53.

Cooper, Franklin S. 1940. "Location and Extent of Industrial Research Activity in the United States." In *Research - A National Resource. II. Industrial Research*, edited by National Research Council, 173–187. Washington, DC: United States Government Printing Office.

Cooren, François. 2004. "Textual Agency: How Texts Do Things in Organizational Settings." *Organization* 11 (3):373–393.

Crane, E.J. 1944. "Growth of Chemical Literature. Contributions of Certain Nations and the Effects of War." *Chemical & Engineering News* 22 (17):1478–1481, 1496.

Crane, E.J., Austin M. Patterson, and Eleanor B. Marr. 1927. *A Guide to the Literature of Chemistry*. New York: John Wiley & Sons.

Creed, W. E. Douglas, Maureen A. Scully, and John R. Austin. 2002. "Clothes Make the Person? The Tailoring of Legitimating Accounts and the Social Construction of Identity." *Organization Science* 13 (5):475–496.

Croidieu, Grégoire, and Philippe Monin. 2009. "Why Effective Entrepreneurial Innovations Sometimes Fail to Diffuse: Identity-Based Interpretations of Appropriateness in the Saint-Émilion, Languedoc, Piedmont, and Golan Heights Wine Regions." In *Institutions and Entrepreneurship*, edited by Wesley D. Sine and Robert J. David, 287–328. Biggleswade: Emerald Group Publishing.

Crosland, Maurice. 2005. "Early Laboratories c.1600-c.1800 and the Location of Experimental Science." *Annals of Science* 62 (2):233–253.

Czarniawska, Barbara, and Bernward Joerges. 1996. "Travels of Ideas." In *Translating Organizational Change*, edited by Barbara Czarniawska and Guje Sevón, 13–48. Berlin/New York: de Gruyter.

Czarniawska, Barbara, and Guje Sevón, eds. 2003. *The Northern Lights. Organization Theory in Scandinavia*. Copenhagen: Liber & Abstrakt & Copenhagen Business School Press.

Czarniawska, Barbara, and Guje Sevón. 2005. "Translation Is a Vehicle, Imitation Its Motor, and Fashion Sits at the Wheel." In *Global Ideas. How Ideas, Object and Practices Travel in the Global Economy*, edited by Barbara Czarniawska and Guje Sevón, 7–12. Copenhagen: Liber.

Dalton, Richard F. 1995. *The Life and Legacy of Charles J. Strosacker*. Midland, MI: Northwood University Press.

Daniels, George H. 1971. *Science in American Society. A Social History*. New York: Alfred A. Knopf.

Daniels, George H. 1976. "The Process of Professionalization in American Science: The Emergent Period, 1820-1860." In *Science in America Since 1820*, edited by Nathan Reingold. New York: Science History Publications.

Davis, Gerald F., and Christopher Marquis. 2005. "Prospects for Organization Theory in the Early Twenty-First Century: Institutional Fields and Mechanisms." *Organization Science* 16 (4):332–343.

Davis, Lance E., and Daniel Kevles. 1974. "The National Research Fund: A Case Study in the Industrial Support of Academic Science." *Minerva* 12 (2):207–220.

Deephouse, David L. 1996. "Does Isomorphism Legitimate?" *The Academy of Management Journal* 39 (4):1024–1039.

Deetz, Stanley. 2003. "Reclaiming the Legacy of the Linguistic Turn." *Organization* 10 (3):421–429.

Dennis, Michael Aaron. 1987. "Accounting for Research: New Histories of Corporate Laboratories and the Social History of American Science." *Social Studies of Science* 17:479–518.

DiMaggio, Paul J., and Walter W. Powell. 1983. "The Iron Cage Revisited: Institutional Isomorphism and Collective Rationality in Organizational Fields." *American Sociological Review* 48 (2):147–160.

DiMaggio, Paul J., and Walter W. Powell. 1991. "Introduction." In *The New Institutionalism in Organizational Analysis*, edited by Walter W. Powell and Paul J. DiMaggio. Chicago, IL: University of Chicago Press.

Djelic, Marie-Laure. 2008. "Sociological Studies of Diffusion: Is History Relevant?" *Socio-Economic Review* 6:538–557.

Dow Chemical Company. 1947. *50th Anniversary*. Midland, MI: The Dow Chemical Company.

Dow Chemical Company. 1997. *The Dow Chemical Company. A Legacy of Leadership. The First 100 Years*. Midland, MI: The Dow Chemical Company.

Duncan, Robert Kennedy. 1907a. *The New Knowledge. A Popular Account of the New Physics and the New Chemistry in Their Relation to the New Theory of Matter*. London: Hodder and Stoughton.

Duncan, Robert Kennedy. 1907b. *Chemistry for Commerce*. London/New York: Harper & Brothers.

Duncan, Robert Kennedy. 1909. "Industrial Fellowships." *Journal of Industrial and Engineering Chemistry* 1 (8):600–603.

Dupree, A. Hunter. 1957. *Science in the Federal Government. A History of Policies and Activities to 1940*. Cambridge, MA: The Belknap Press.

Dupree, A. Hunter. 1976. "The National Pattern of American Learned Societies, 1769-1863." In *The Pursuit of Knowledge in the Early American Republic: American Scientific and Learned Societies from Colonial Times to the Civil War*, edited by Alexandra Oleson, 21–32. Baltimore, MD: Johns Hopkins University Press.

Dupree, A. Hunter. 1979. "The National Academy of Sciences and the American Definition of Science." In *The Organization of Knowledge in Modern America, 1860-1920*, edited by Alexandra Oleson and John Voss, 342–363. Baltimore: The Johns Hopkins University Press.

Eastman Kodak Company. 1989. *Journey: 75 Years of Kodak Research*. Rochester, NY: Eastman Kodak Company.

Electrical World. 1913. Industrial Research. *Electrical World* 62 (18): 877.

Elfenbein, Julien. 1960. *Business Journalism*. New York: Greenwood Press.

Engwall, Lars. 2003. "On the Origin of the Northern Lights." In *The Northern Lights. Organization Theory in Scandinavia*, edited by Barbara Czarniawska and Guje Sevón, 395–412. Copenhagen: Liber & Abstrakt & Copenhagen Business School Press.

Fagerberg, Jan. 2005. "Innovation. A Guide to the Literature." In *The Oxford Handbook of Innovation*, edited by Jan Fagerberg, David C. Mowery, and Richard R. Nelson. Oxford: Oxford University Press.

Fairclough, Norman. 1992. *Discourse and Social Change*. Cambridge, UK: Polity Press.

Fennell, Roger. 1994. *History of IUPAC, 1919-1987*. Oxford: Blackwell Science.

Fisk, Catherine L. 2009. *Working Knowledge. Employee Innovation and the Rise of Corporate Intellectual Property, 1800-1930*. Chapel Hill, NC: The University of North Carolina Press.

Fleming, A.P.M. 1917. *Industrial Research in the United States of America*. London: Department of Scientific and Industrial Research by His Majesty's Stationery Office.

Flick, Uwe. 2007. *Managing Quality in Qualitative Research*. London: SAGE.

Fligstein, Neil. 1990. *The Transformation of Corporate Control*. Cambridge, MA/London, England: Harvard University Press.

Fligstein, Neil. 2008. "Chandler and the Sociology of Organizations." *Business History Review* 82 (2):241–250.

Foucault, Michel. 1981. *Archäologie des Wissens*. Frankfurt am Main: Suhrkamp.

Freyer, Tony. 1995. "Legal Restraints on Economic Coordination: Antitrust in Great Britain and America, 1880-1920." In *Coordination and Information. Historical Perspectives on the Organization of Enterprise*, edited by Naomi Lamoreaux and Daniel M.G. Raff. Chicago/London: University of Chicago Press.

Friedel, Robert. 1992. "Perspiration in Perspective: Changing Perceptions of Genius and Expertise in American Invention." In *Inventive Minds. Creativity in Technology*, edited by Robert J. Weber and David N. Perkins, 11–26. New York/Oxford: Oxford University Press.

Furman, Jeffrey L., and Megan J. MacGarvie. 2007. "Academic Science and the Birth of Industrial Research Laboratories in the U.S. Pharmaceutical Industry." *Journal of Economic Behavior & Organization* 63:756–776.

Furter, William F. 1980b. "Chemical Engineering and the Public Image." In *History of Chemical Engineering*, edited by William F. Furter. Washington, DC: American Chemical Society.

Fyfe, Aileen. 2016. "Journals and Periodicals." In *A Companion to the History of Science*, edited by Bernard Lightman. West Sussex, UK: John Wiley & Sons, Ltd.

Galambos, Louis. 1975. *The Public Image of Big Business in America, 1880-1940. A Quantitative Study in Social Change*. Baltimore/London: The Johns Hopkins University Press.

Galambos, Louis. 1979. "The American Economy and the Reorganization of the Sources of Knowledge." In *The Organization of Knowledge in Modern*

America, 1860-1920, edited by Alexandra Oleson and John Voss, 269–282. London/Baltimore: Johns Hopkins University Press.

Galambos, Louis. 1984. "Technology, Political Economy, and Professionalization: Central Themes of the Organizational Synthesis." *Business History Review* 57 (4):471–493.

Galambos, Louis. 2010. "The Role of Professionals in the Chandler Paradigm." *Industrial and Corporate Change* 19 (2):377–398.

Galambos, Louis, and Jane E. Sewell. 1995. *Networks of Innovation: Vaccine Development at Merck, Sharp & Dohme, and Mulford, 1895-1995*. New York: Cambridge University Press.

Garcia-Belmar, Antonio. 2014. "Sites of Chemistry in the Nineteenth Century." *Ambix* 61 (2):109–114.

Geiger, Roger L. 1998. "The Rise and Fall of Useful Knowledge: Higher Education for Science, Agriculture & the Mechanic Arts, 1850-1875." *History of Higher Education* 18:47–66.

General Chemical Company. 1919. *The General Chemical Company after Twenty Years, 1899-1919*. New York: Sackett & Wilhelms Corporation.

Georgaca, Eugenie, and Evrinomy Avdi. 2011. "Discourse Analysis." In *Qualitative Research Methods in Mental Health and Psychotherapy: A Guide for Students and Practitioners*, edited by David Harper and Andrew R. Thompson, 147–162. West Sussex, UK: Wiley-Blackwell.

Gertner, Jon. 2012. *The Idea Factory. Bell Labs and the Great Age of American Innovation*. London: Penguin Books.

Gibb, George Sweet, and Evelyn H. Knowlton. 1976. *History of Standard Oil Company (New Jersey): The Resurgent Years 1911-1927*. New York: Arno Press.

Gieryn, Thomas F. 1983. "Boundary-Work and the Demarcation of Science from Non-Science: Strains and Interests in Professional Ideologies of Scientists." *American Sociological Review* 48 (6):781–795.

Gieryn, Thomas F. 1995. "Boundaries of Science." In *Handbook of Science and Technology Studies*, edited by Sheila Jasanoff, Gerald E. Markle, James C. Petersen, and Trevor Pinch, 393–443. Thousand Oaks: SAGE.

Gieryn, Thomas F. 1999. *Cultural Boundaries of Science. Credibility on the Line*. Chicago/London: University of Chicago Press.

Godin, Benoit. 2003. "Measuring Science: Is There "Basic Research" without Statistics?" *Social Science Information* 42(1): 57–90.

Godin, Benoît. 2006. "The Linear Model of Innovation: The Historical Construction of an Analytical Framework." *Science Technology Human Values* 31 (6):639–667.

Godin, Benoît, and Joseph Lane. 2011. "Research or Development? A Short History of Research and Development as Categories." *Gegenworte 26*: 45–48.

Godin, Benoît, and Désirée Schauz. 2016. "The Changing Identity of Research: A Cultural and Conceptual History." *History of Science*:1–31. doi: 10.1177/0073275316656007

Gooday, Graeme. 1998. "The Premisses of Premises: Spatial Issues in the Historical Construction of Laboratory Credibility." In *Making Space for Science. Territorial Themes in the Shaping of Knowledge*, edited by Crosbie Smith and Jon Agar. London: MacMillan Press Ltd.

Gooday, Graeme. 2008. "Placing or Replacing the Laboratory in the History of Science?" *Isis* 99 (4):783–795.

Gortler, Leon. 2000. "Merck in America: The First 70 Years from Fine Chemicals to Pharmaceutical Giant." *Bulletin for the History of Chemistry* 25 (1):1–9.

Graham, Margaret B.W. 1986. *RCA and the VideoDisc: The Business of Research*. Cambridge: Cambridge University Press.

Graham, Margaret B.W., and Bettye H. Pruitt. 1990. *R&D for Industry. A Century of Technical Innovation at Alcoa.* Cambridge: Cambridge University Press.

Grant, David, Cynthia Hardy, Cliff Oswick, and Linda Putman. 2004. "Introduction: Organizational Discourse: Exploring the Field." In *The SAGE Handbook of Organizational Discourse*, edited by David Grant, Cynthia Hardy, Cliff Oswick, and Linda Putman. London: SAGE.

Grant, David, Cynthia Hardy, Cliff Oswick, and Linda Putnam, eds. 2004. *The SAGE Handbook of Organizational Discourse*. London: SAGE.

Grant, David, and Robert J. Marshak. 2011. "Toward a Discourse-Centered Understanding of Organizational Change." *The Journal of Applied Behavioral Science* 47 (2):204–235.

Grant, David, Grant Michelson, Cliff Oswick, and Nick Wailes. 2005. "Guest Editorial: Discourse and Organizational Change." *Journal of Organizational Change Management* 18 (1):6–15.

Green, Sandy Edward Jr., and Yuan Li. 2011. "Rhetorical Institutionalism: Language, Agency, and Structure in Institutional Theory since Alvesson 1993." *Journal of Management Studies* 48 (7):1662–1697.

Green, Sandy Edward Jr., Yuan Li, and Nitin Nohria. 2009. "Suspended in Self-Spun Webs of Significance: A Rhetorical Model of Institutionalization and Institutionally Embedded Agency." *Academy of Management Journal* 52 (1):11–36.

Greenwood, Royston, and C.R. Hinings. 1996. "Understanding Radical Organizational Change: Bringing together the Old and the New Institutionalism." *Academy of Management Review* 21 (4):1022–1054.

Greenwood, Royston, Christine Oliver, Kerstin Sahlin, and Roy Suddaby. 2008. "Introduction." In *The SAGE Handbook of Organizational Institutionalism*, edited by Royston Greenwood, Christine Oliver, Kerstin Sahlin, and Roy Suddaby, 1–46. London: SAGE.

Haber, Samuel. 1964. *Efficiency and Uplift. Scientific Management in the Progressive Era 1890-1920*. Chicago: The University of Chicago Press.

Hall, Courtney Robert. 1954. *History of American Industrial Science*. New York: Library Publishers.

Hall, Peter A., and Rosemary C.R. Taylor. 1996. "Political Science and the Three New Institutionalisms." *Political Studies* 44:936–957.

Hardy, Cynthia. 2001. "Researching Organizational Discourse." *International Studies of Management & Organization* 31 (3):25–47.

Hardy, Cynthia. 2004. "Scaling Up and Bearing Down in Discourse Analysis: Questions Regarding Textual Agencies and Their Context." *Organization* 11 (3):415–425.

Hardy, Cynthia, Thomas B. Lawrence, and David Grant. 2005. "Discourse and Collaboration: The Role of Conversation and Collective Identity." *Academy of Management Review* 30 (1):58–77.

Hawkins, Hugh. 1979. "University Identity: The Teaching and Research Functions." In *The Organization of Knowledge in Modern America, 1860-1920*, edited by Alexandra Oleson and John Voss, 285–312. London/ Baltimore: Johns Hopkins University Press.

Hawkins, Laurence A. 1950. *Adventure into the Unknown. The First Fifty Years of the General Electric Research Laboratory*. New York: William Morrow & Co.

Haynes, Williams. 1939. *Chemical Pioneers: The Founders of the American Chemical Industry*. New York: D. Van Nostrand Company.

Haynes, Williams. 1954a. *American Chemical Industry. Background and Beginnings*, Vol. I. New York: D. Van Nostrand Company, Inc.

Haynes, Williams. 1954b. *American Chemical Industry. The World War I Period: 1912-1922*. New York: D. van Nostrand Company, Inc.

Haynes, Williams. 1954c. *American Chemical Industry. Decade of New Products*. New York: D. van Nostrandt Company, Inc.

Hays, Samuel P. 1959. *Conservation and the Gospel of Efficiency*. Cambridge, MA: Harvard University Press.

Hendrick, Elwood. 1917. *Everyman's Chemistry. The Chemist's Point of View and His Recent Work Told for the Layman*. London/New York: Harper & Brothers.

Heracleous, Loizos, and Michael Barrett. 2001. "Organizational Change as Discourse: Communicative Actions and Deep Structures in the Context of Information Technology Implementation." *The Academy of Management Journal* 44 (4):755–778.

Hidy, Ralph W., and Muriel E. Hidy 1995. *Pioneering in Big Business 1882-1911*. New York: Harper & Brothers.

Hintz, Eric S. 2011. "The Post-Heroic Generation: American Independent Inventors, 1900–1950." *Enterprise & Society* 12 (4):732–748.

Hintz, Eric S. 2013. ""Selling the Research Idea": The Genesis of the Industrial Research Institute, 1916-1945." *Research-Technology Management* 2013 (6):46–50.

Hirsch, Paul M. 1986. "From Ambushes to Golden Parachutes: Corporate Takeovers as an Instance of Cultural Framing and Institutional Integration." *American Journal of Sociology* 91 (4):800–837.

Hoddeson, Lillian. 1977. "The Roots of Solid State Research at Bell Labs." *Physics Today* 30 (3): 23–31.

Hoddeson, Lillian. 1980. "The Entry of Quantum Theory of Solids into the Bell Telephone Laboratories, 1925-40: A Case-Study of the Industrial Application of Fundamental Science." *Minerva* 18 (3):422–447.

Hoddeson, Lillian. 1981. "The Emergence of Basic Research in the Bell Telephone System, 1875-1915." *Technology and Culture* 22 (3):512–544.

Hoffman, Andrew J. 1999. "Institutional Evolution and Change: Environmentalism and the U.S. Chemical Industry." *Academy of Management Journal* 42 (4):351–371.

Holland, Maurice, and Henry F. Pringle. 1928. *Industrial Explorers*. London/ New York: Harper & Brothers.

Homburg, Ernst. 1992. "The Emergence of Research Laboratories in the Dyestuffs Industry, 1870-1900." *British Journal for the History of Science* 25 (1):91–111.

Hoover, Herbert. 1926. "The Vital Need for Greater Financial Supprot to Pure Science Research." *Mechanical Engineering* 48:6–8.

Hounshell, David. 1984. "Science and Invention." *Annals of the New York Academy of Science* 424:183–192.

Hounshell, David. 1996. "The Evolution of Industrial Research in the United States." In *Engines of Innovation. U.S. Industrial Research at the End of an Era*, edited by Richard S. Rosenbloom and William J. Spencer, 13–85. Boston, MA: Harvard Business School Press.

Hounshell, David, and John Kenly Smith Jr. 1988. *Science and Corporate Strategy. Du Pont R&D, 1902-1980*. Cambridge: Cambridge University Press.

Hughes, Thomas P. 1989. *American Genesis: A Century of Invention and Technological Enthusiasm, 1870-1970*. New York: Viking.

Hyland, Ken. 2011. "Academic Discourse." In *The Bloomsbury Companion to Discourse Analysis*, edited by Ken Hyland and Brian Paltridge, 171–184. London: Bloomsbury.

Hyland, Ken, and Brian Paltridge, eds. 2011. *The Bloomsbury Companion to Discourse Analysis*. London: Bloomsbury.

Iedema, Rick. 2011. "Discourse Studies in the 21st Century: A Response to Mats Alvesson and Dan Kärreman's 'Decolonizing Discourse'." *Human Relations* 64 (9):1121–1146.

Ihde, Aaron J. 1964. *The Development of Modern Chemistry*. New York: Harper & Row.

Israel, Paul. 1992. *From Machine Shop to Industrial Laboratory. Telegraphy and the Changing Context of American Invention, 1830-1920*. Baltimore & London: The Johns Hopkins University Press.

Israel, Paul. 2002. "Inventing Industrial Research: Thomas Edison and the Menlo Park Laboratory." *Endeavour* 26 (2):48–54.

Jackson, Catherine M. 2016. "The Laboratory." In *A Companion to the History of Science*, edited by Bernard Lightman, 296–309. West Sussex, UK: John Wiley & Sons, Ltd.

Jäger, Siegfried. 2001. "Discourse and Knowledge: Theoretical and Methodical Aspects of a Critical Discourse and Dispositive Analysis." In *Methods of Critical Discourse Analysis*, edited by Ruth Wodak and Michael Meyer, 32–62. London: SAGE Publications.

James, Frank A.J.L., ed. 1989. *The Development of the Laboratory. Essays on the Place of Experiment in Industrial Civilization*. London: Macmillan.

James, T.H. 1990. A *Biography-Autobiography of Charles Edward Kenneth Mees, Pioneer of Industrial Research*. Rochester, NY: Photographic Research Laboratories, Eastman Kodak Company.

Jenkins, Reese V. 1975. *Images and Enterprise. Technology and the American Photographic Industry 1839 to 1925*. Baltimore/London: Johns Hopkins University Press.

Jonsson, Stefan. 2009. "Refraining from Imitation: Professional Resistance and Limited Diffusion in a Financial Market." *Organization Science* 20 (1):172–186.

Kahn Jr., E.J. 1986. *The Problem Solvers. A History of Arthur D. Little, Inc.* Boston/Toronto: Little, Brown and Company.

Kaldewey, David, and Désirée Schauz. 2018. *Basic and Applied Research. The Language of Science Policy in the Twentieth Century*. New York: Berghahn.

Karpiuk, Robert S. 1984. *Dow Research Pioneers: Recollections*. Midland, MI: Pendell Publishing Company.

Katz, Elihu. 1999. "Theorizing Diffusion: Tarde and Sorokin Revisited." *The Annals of the American Academy of Political and Social SCience* 566 (1):144–155.

Katz, Elihu, Herbert Hamilton, and Martin L. Levin. 1963. "Traditions of Research on the Diffusion of Innovation." *American Sociological Review* 28 (2):237–252.

Keenoy, Tom, and Cliff Oswick. 2004. "Organizing Textscapes." *Organization Studies* 25 (1):135–142.

Keller, Reiner. 2010. "Der Müll der Gesellschaft. Eine wissenssoziologische Diskursanalyse." In *Handbuch Sozialwissenschaftliche Diskursanalyse. Band 2: Forschungspraxis*, edited by Reiner Keller, Andreas Hirseland, Werner Schneider, and Willy Viehöver, 197–232. Wiesbaden: VS.

Keller, Reiner, Andreas Hirseland, Werner Schneider, and Willy Viehöver, eds. 2010. *Handbuch Sozialwissenschaftliche Diskursanalyse. Band 2: Forschungspraxis*, 4th ed. Wiesbaden: VS.

Kevles, Daniel. 1968. "George Ellery Hale, the First World War, and the Advancement of Science in America." *Isis* 59 (4):427–437.

Kevles, Daniel. 1971. "Federal Legislation for Engineering Experiment Stations: The Episode of World War I." *Technology and Culture* 12 (2):182–189.

Kevles, Daniel J. 1972. "On the Moral Dilemmas of the American Chemist: Some Notes from History" In *Science, Scientists and Society*, edited by William Beranek Jr., 2–16. Tarrytown-on-Hudson, NY: Bogden & Quigley, Inc.

Kevles, Daniel. 1979. "The Physics, Mathematics, and Chemistry Communities: A Comparative Analysis." In *The Organization of Knowledge in Modern America, 1860-1920*, edited by Alexandra Oleson and John Voss, 139–172. Baltimore: The Johns Hopkins University Press.

Kevles, Daniel. 1995. *The Physicists. The History of a Scientific Community in Modern America*. Cambridge/London: Harvard Unviersity Press.

Kevles, Daniel. 2013. "Not a Hundred Millionaires: The National Academy and the Expansion of Federal Science in the Gilded Age." *Issues in Science and Technology* 29 (2):37–46.

Klein, Ursula. 2008. "The Laboratory Challenge. Some Revisions of the Standard View of Early Modern Experimentation." *Isis* 99 (4):769–782.

Klincewicz, Krzysztof. 2006. *Management Fashions. Turning Best-Selling Ideas into Objects and Institutions*. New Brunswick/London: Transaction Publishers.

Kline, Ronald. 1995. "Construing "Technology" as "Applied Science": Public Rhetoric of Scientists and Engineers in the United States, 1880-1945." *Isis* 86 (2):194–221.

Kline, Ronald, and Thomas C. Lassman. 2005. "Competing Research Traditions in American Industry: Uncertain Alliances between Science and Engineering at Westinghouse Electric, 1886-1935." *Enterprise & Society* 6 (4):601–645.

Knorr-Cetina, Karin D. 1992. "The Couch, the Cathedral, and the Laboratory: On the Relationship between Experiment and Laboratory in Science." In *Science as Practice and Culture*, edited by Andrew Pickering, 113–138. Chicago/London: The University of Chicago Press.

Kohler, Robert E. 1990. "The Ph.D. Machine: Building on the Collegiate Base." *Isis* 81 (4):638–662.

Kohler, Robert E. 2002. "Labscapes: Naturalizing the Lab." *History of Science* 40 (4):473–501.

Kohler, Robert E. 2008. "Lab History: Reflections." *Isis* 99 (4):761–768.

LaFollette, Marcel C. 1990. *Making Science Our Own. Public Images of Science 1910-1955*. London/Chicago: University of Chicago Press.

Lamnek, Siegfried, and Claudia Krell. 2016. *Qualitative Sozialforschung*. Weinheim/Basel: Beltz.

Lamoreaux, Naomi. 1985. *The great merger movement in American business, 1895-1904*. Cambridge: Cambridge University Press.

Lamoreaux, Naomi, Daniel M.G. Raff, and Peter Temin. 2003. "Beyond Markets and Hierarchies: Toward a New Synthesis of American Business History." *The American Historical Review* 108 (2):404–433.

Lamoreaux, Naomi, and Kenneth Sokoloff. 1999. "Inventors, Firms, and the Market for Technology in the Late Nineteenth and Early Twentieth Century." In *Learning by Doing in Markets, Firms, and Countries*, edited by Naomi Lamoreaux, Daniel M.G. Raff and Peter Temin, 19–60. Chicago/London: University of Chicago Press.

Lamoreaux, Naomi, and Kenneth Sokoloff. 2009. "The Rise and Decline of the Independent Inventor: A Schumpeterian Story?" In *The Challenge of Remaining Innovative. Insights from Twentieth-Century American Business*, edited by Sally H. Clarke, Naomi Lamoreaux, and Steven W. Usselman, 43–78. Stanford, California: Stanford University Press.

Laporte, Deirdre. 1983. "Bell Laboratories: The Beginnings of Scientific Research in an Industrial Setting." *Annals of the New York Academy of Science* 412:85–100.

Larson, Henrietta M. 1969. "Contours of Change: Standard Oil Company (New Jersey), 1882-1950." *Nebraska Journal of Economics and Business* 8 (3):3–19.

Latour, Bruno. 1983. "Give Me a Laboratory and I Will Raise the World." In *Science Observed. Perspectives on the Social Study of Science*, edited by Karin D. Knorr-Cetina and Michael Mulkay, 141–170. London: Sage.

Latour, Bruno. 1986. "The Powers of Association." In *Power, Action, and Belief. A New Sociology of Knowledge?*, edited by John Law, 261–277. London: Routledge.

Lawrence, Thomas B., and Nelson Phillips. 2004. "From Moby Dick to Free Willy: Macro-Cultural Discourse and Institutional Entrepreneurship in Emerging Institutional Fields." *Organization* 11 (5):689–711.

Lazby, Clarence G. 1966. "Science and the Military." In *Science and Society in the United States*, edited by David D. Van Tassel and Michael G. Hall. Homewood, IL: The Dorsey Press.

Leblebici, Huseyin, Gerald R. Salancik, Anne Copay, and Tom King. 1991. "Institutional Change and the Transformation of Interorganizational Fields:

An Organizational History of the U.S. Radio Broadcasting Industry." *Administrative Science Quarterly* 36 (3):333–363.

Levenstein, Margaret. 1998. *Accounting for Growth. Information Systems and the Creation of the Large Corporation.* Stanford, CA: Stanford University Press.

Liebenau, Jonathan. 1985. "Innovation in Pharmaceuticals: Industrial R&D in the Early Twentieth Century." *Research Policy* 14:179–187.

Lipartito, Kenneth. 2009. "Rethinking the Invention Factory: Bell Laboratories in Perspective." In *The Challenge of Remaining Innovative. Insights from Twentieth-Century American Business*, edited by Sally H. Clarke, Naomi Lamoreaux, and Steven W. Usselman, 132–159. Stanford, CA: Stanford University Press.

Lovelace, B.F., and Charles C. Thomas. 1919. *Annual Chemical Directory of the United States*, 2nd ed. Baltimore, MD: Williams & Wilkins Company.

Lovelace, B.F., and Charles C. Thomas. 1920. *Annual Chemical Directory of the United States*, 3rd ed. Baltimore, MD: Williams & Wilkins Company.

Lucier, Paul. 2009. "The Professional and the Scientist in Nineteenth-Century America." *Isis* 100 (4):699–732.

Lucier, Paul. 2012. "The Origins of Pure and Applied Science in Gilded Age America." *Isis* 103:527–536.

Lucier, Paul. 2016. "Commercial Science." In *A Companion to the History of Science*, edited by Bernard Lightman, 268–281. West Sussex, UK: John Wiley & Sons, Ltd.

Machin, David, and Andrea Mayr. 2012. *How To Do Critical Discourse Analysis. A Multimodal Introduction.* Los Angeles: SAGE.

Maguire, Steve, and Cynthia Hardy. 2006. "The Emergence of New Global Institutions: A Discursive Perspective." *Organization Studies* 27 (1):7–29.

Maguire, Steve, and Cynthia Hardy. 2009. "Discourse and Deinstitutionalization: The Decline of DDT." *Academy of Management Journal* 52 (1):148–178.

Maguire, Steve, and Cynthia Hardy. 2013. "Organizing Processes and the Construction of Risk: A Discursive Approach." *Academy of Management Journal* 56 (1):231–255.

Mahajan, Vijay, and Robert A. Peterson. 1985. *Models for Innovation Diffusion.* London: Sage Publications.

Manufacturing Chemists' Association, Inc. 1972. *MCA 1872-1972. A Centennial History.* Washington, DC: Manfucaturing Chemists's Association.

McCormick, G. S. 1921. "Who Is a Chemist?" *Chemical and Metallurgical Engineering* 24 (10):416.

Mees, C. E. Kenneth. 1920. The Organizat*ion of Industrial Scientific Research.* New York: McGraw-Hill Book Company.

Mees, C. E. Kenneth. 1956. *An Address to the Senior Staff of the Kodak Research Laboratories, November 9, 1955.* Rochester, NY: Kodak Research Laboratories.

Meyer, John W. 1996. "Otherhood: The Promulgation and Transmission of Ideas in the Modern Organizational Environment." In *Translating Organizational Change*, edited by Barbara Czarniawska and Guje Sevón, 241–252. Berlin: de Gruyter.

Meyer, John W., and Brian Rowan. 1977. "Institutionalized Organizations: Formal Structure as Myth and Ceremony." *American Journal of Sociology* 83 (2):340–363.

Meyer-Thurow, Georg. 1982. "The Industrialization of Invention: A Case Study gfrom the German Chemical Industry." *Isis* 73 (3):363–381.

Mica, Adriana. 2013. "From Diffusion to Translation and Back. Disembedding - Re-Embedding and Re-Invention in Sociological Studies of Diffusion." *Polish Sociological Review* 181 (1):3–19.

Miller, David Philip. 2011. "The Paradoxes of Patenting at General Electric. Isador Ladoff's Journey from Siberian Exile to the Heart of Corporate Capitalism." *Isis* 102 (4):634–658.

Mirowski, Philip, and Esther-Mirjam Sent. 2008. "The Commercialization of Science and the Response of STS." In *The Handbook of Science and Technology Studies*, edited by Edward J. Hackett, Olga Amsterdamska, Michael Lynch, and Judy Wacjman. Cambridge, MA/London, England: MIT Press.

Mizruchi, Mark S., and Lisa C. Fein. 1999. "The Social Construction of Organizational Knowledge: A Study of the Uses of Coercive, Mimetic, and Normative Isomorphism." *Administrative Science Quarterly* 44 (4):653–683.

Mody, Cyrus C.M. 2011. *Instrumental Community. Probe Microscopy and the Path to Nanotechnology*. Cambridge, MA/London, England: MIT Press.

Mody, Cyrus C.M. 2016a. "Fabricating an Organizational Field for Research: US Academic Microfabrication Facilities in the 1970s and 1980s." In *Innovation in Science and Organizational Renewal*, edited by Thomas Heinze and Richard Münch, 21–52. New York: Palgrave.

Mody, Cyrus C.M. 2016b. "The Professional Scientist." In *A Companion to the History of Science*, edited by Bernard Lightman. West Sussex, UK: John Wiley & Sons, Ltd.

Mott, Frank Luther. 1957a. *A History of American Magazines 1741-1850*. Cambridge, MA: Harvard University Press.

Mott, Frank Luther. 1957b. *A History of American Magazines 1885-1905*. Cambridge, MA: Harvard University Press.

Mott, Frank Luther. 1970a. *A History of American Magazines 1850-1865*. Cambridge, MA: Harvard University Press.

Mott, Frank Luther. 1970b. *A History of American Magazines 1865-1885*. Cambridge, MA: Harvard University Press.

Mowery, David C. 1981. *The Emergence and Growth of Industrial Research in American Manufacturing, 1899-1945*. Ann Arbor, MI/London, England: University Microfilms International.

Mowery, David C. 1983a. "The Relationship Between Intrafirm and Contractual Forms of Industrial Research in American Manufacturing, 1900-1940." *Explorations in Economic History* 20:351–374.

Mowery, David C. 1983b. "Industrial Research and Firm Size, Survival, and Growth in American Manufacturing, 1921-1946: An Assessment." *The Journal of Economic History* 43 (4):953–980.

Mowery, David C. 1984. "Firm Structure, Government Policy, and the Organization of Industrial Research: Great Britain and the United States, 1900-1950." *Business History Review* 58:504–531.

Mowery, David C. 1995. "The Boundaries of the U.S. Firm in R&D." In *Coordination and Information. Historical Perspectives on the Organization of Enterprise*, edited by Naomi Lamoreaux and Daniel M.G. Raff, 147–182. Chicago/London: University of Chicago Press.

Mowery, David C., and Nathan Rosenberg. 1989. *Technology and the Pursuit of Economic Growth*. New York: Cambridge University Press.

Mowery, David C., and David J. Teece. 1996. "Strategic Alliances And Industrial Research." In *Engines of Innovation. U.S. Industrial Research at the End of an Era*, edited by Richard S. Rosenbloom and William J. Spencer, 111–129. Boston, Massachusetts: Harvard Business School Press.

Munir, Kamal A., and Nelson Phillips. 2005. "The Birth of the 'Kodak Moment': Institutional Entrepreneurship and the Adoption of New Technologies." *Organization Studies* 26 (11):1665–1687.

Murmann, Johann Peter. 2000. "Knowledge and Competitive Advantage in the Synthetic Dye Industry, 1850-1914: The Coevolution of Firms, Technology, and National Institutions in Great Britain, Germany, and the United States." *Enterprise & Society* 1:699–704.

Murmann, Johann Peter. 2002. "Chemical Industries after 1850." In *The Oxford Encyclopedia of Economic History*, edited by Joel Moykr and Oxford University Press. Oxford: Oxford University Press.

Murmann, Johann Peter, and Ernst Homburg. 2001. "Comparing Evolutionary Dynamics across Different National Settings: The Case of the Synthetic Dye Industry, 1857–1914." *Journal of Evolutionary Economics* 11:177–205.

Näsie, Salme, and Carsten Rohde. 2007. "Development of Cost and Management Accounting Ideas in the Nordic Countries." In *Handbook of Management Accounting Research*, edited by Christopher S. Chapman, Anthony G. Hopwood, and Michael D. Shields, 1091–1118. Amsterdam: Elsevier.

National Academy of Sciences. 1971. *Scientific, Technical and Related Societies of the United States*. Washington, DC: National Academy of Sciences.

National Research Council. 1920. *Research Laboratories in Industrial Establishments of the United States of America*, Vol. I, Part 2, Bulletin of the National Research Council. Washington, DC: National Research Council, National Academy of Sciences.

National Research Council. 1921. *Research Laboratories in Industrial Establishments of the United States, Including Consulting Laboratories, Bulletin of the National Research Council*. Washington, DC: National Research Council, National Academy of Sciences.

National Research Council. 1927. *Industrial Research Laboratories of the United States*, *Bulletin of the National Research Council*. Washington, DC: National Research Council, National Academy of Sciences.

National Research Council. 1931. *Industrial Research Laboratories of the United States*, *Bulletin of the National Research Council*. Washington, DC: National Research Council, National Academy of Sciences.

National Research Council. 1933. *Industrial Research Laboratories of the United States*, *Bulletin of the National Research Council*. Washington, DC: National Research Council, National Academy of Sciences.

National Research Council. 1938. *Industrial Research Laboratories of the United States, Bulletin of the National Research Council*. Washington, DC: National Research Council, National Academy of Sciences.

National Research Council. 1940a. *Industrial Research Laboratories of the United States, Bulletin of the National Research Council*. Washington, DC: National Research Council, National Academy of Sciences.

National Research Council. 1940b. *Research - A National Resource. II. Industrial Research*. Washington, DC: United States Government Printing Office.

National Research Council. 1946. *Industrial Research Laboratories of the United States. Including Consulting Research Laboratories*, 8th ed. Washington, DC: National Research Council/National Academy of Sciences.

Nee, Victor. 1998. "Sources of New Institutionalism." In *The New Institutionalism in Sociology*, edited by Mary C. Brinton and Victor Nee, 1–16. New York: Russel Sage Foundation.

Nicholas, Tom. 2010. "The Role of Independent Invention in U.S. Technological Development, 1880–1930." *The Journal of Economic History* 70 (1):57–82.

Nielsen, Anita Kildebaek, and Soňa Štrbáňová. 2008. *Creating Networks in Chemistry. The Founding and Early History of Chemical Societies in Europe*. Cambridge: The Royal Society of Chemistry.

Noble, David. 1974. *Science and Technology in the Corporate Search for Order: American Engineers and Social Reform, 1900-1929*, PhD Dissertation. Rochester, NY: University of Rochester.

Noyes, W. A. 1908. "Chemical Publications in America in Relation to Chemical Industry." *Science* 28 (712): 225–227.

Nummedal, Tara. 2016. "The Alchemist." In *A Companion to the History of Science*, edited by Bernard Lightman, 58–70. West Sussex, UK: John Wiley & Sons, Ltd.

Nutting, P.G. 1917. "Organized Knowledge and National Welfare." *Science* 46 (1185):247–252.

Nye, David E. 1985. *Image Worlds. Corporate Identities at General Electric, 1890-1930*. Cambridge, MA: The MIT Press.

Ophir, Adi, and Steven Shapin. 1991. "The Place of Knowlede: A Methodological Survey." *Science in Context* 4 (1):3–21.

Owens, Larry. 1997. "Science in the United States." In *Science in the Twentieth Century*, edited by John Krige and Dominique Pestre, 821–837. Amsterdam: Harwood Academic Publishers.

Palloni, Alberto. 2001. "Diffusion in Sociological Analysis." In *Diffusion Processes and Fertility Transition*, edited by John B. Casterline, 66–114. Washington DC: National Academy Press.

Perazich, George, and Philip M. Field. 1940. *Industrial Research and Changing Technology*. Philadelphia, PA: Works Project Administration, National Research Project.

Phillips, Nelson, and Cynthia Hardy. 2002. *Discourse Analysis. Investigating Processes of Social Construction*. Thousand Oaks/London/New Delhi: Sage Publications.

Phillips, Nelson, Thomas B. Lawrence, and Cynthia Hardy. 2004. "Discourse and Institutions." *Academy of Management Review* 29 (4):635–652.

Phillips, Nelson, and Namrata Malhotra. 2008. "Taking Social Construction Seriously: Extending the Discursive Approach in Institutional Theory." In *The SAGE Handbook of Organizational Institutionalism*, edited by Royston Greenwood, Christine Oliver, Roy Suddaby and Kerstin Sahlin, 702–720. London: SAGE.

Pielke, Roger Jr. 2012. "Basic Research as a Political Symbol." *Minerva* 50 (3):339–361.

Potter, Jonathan, and Margaret Wetherell. 1987. *Discourse and Social Psychology. Beyond Attitudes and Behaviour*. London: SAGE.

Pursell Jr., Caroll W. 1966. "Science and Government Agencies." In *Science and Society in the United States*, edited by David D. Van Tassel and Michael G. Hall. Homewood, IL: The Dorsey Press.

Pynchon, Thomas. 2006. *Against the Day*. London: Jonathan Cape.

Rae, John. 1979. "The Application of Science to Industry." In *The Organization of Knowledge in Modern America, 1860-1920*, edited by Alexandra Oleson and John Voss, 249–268. London/Baltimore: Johns Hopkins University Press.

Reese, Kenneth M. 1976a. *A Century of Chemistry. The Role of Chemists and the American Chemical Society*. Washington, DC: American Chemical Society.

Reese, Kenneth M. 1976b. "American Chemical Society: The First 100 Years." *Chemical & Engineering News* 54 (15):22–31.

Reich, Leonard S. 1977a. *Radio Electronics and the Development of Industrial Research in the Bell System*. Ann Arbor, MI/London, England: University Microfilms International.

Reich, Leonard S. 1977b. "Research, Patents, and the Struggle to Control Radio: A Study of Big Business and the Uses of Industrial Research." *Business History Review* 51 (2):208–235.

Reich, Leonard S. 1980. "Industrial Research and the Pursuit of Corporate Security: The Early Years of Bell Labs." *Business History Review* 54 (4):504–529.

Reich, Leonard S. 1983. "Irving Langmuir and the Pursuit of Science and Technology in the Corporate Environment." *Technology and Culture* 24 (2):199–221.

Reich, Leonard S. 1985. *The Making of American Industrial Research: Science and Business at GE and Bell, 1876-1926*. Cambridge, UK: Cambridge University Press.

Reich, Leonard S. 1987. "Edison, Coolidge, and Langmuir: Evolving Approaches to American Industrial Research." *The Journal of Economic History* 47 (4):341–351.

Reich, Leonard S. 1992. "Lighting the Path to Profit: GE's Control of the Electric Lamp Industry, 1892-1941." *Business History Review* 66:305–334.

Reingold, Nathan. 1964. *Science in Nineteenth-Century America. A Documentary History*. Chicago: The University of Chicago Press.

Reynolds, Terry S. 1983. *75 Years of Progress - A History of the American Institute of Chemical Engineers 1908-1983*. New York: American Institute of Chemical Engineers.

Reynolds, Terry S. 1991a. "The Engineer in 19th-Century America." In *The Engineer in America. A Historical Anthology from Technology and Culture*, edited by Terry S. Reynolds, 7–26. Chicago: The University of Chicago Press.

Reynolds, Terry S. 1991b. "The Engineer in 20th-Century America." In *The Engineer in America. A Historical Anthology from Technology and Culture*, edited by Terry S. Reynolds, 169–190. Chicago: The University of Chicago Press.

Rhees, David J. 1987. *The Chemists' Crusade: The Rise of an Industrial Science in Modern America, 1907-1922*. Ann Arbor, MI: UMI Dissertation Services.

Rhees, David J. 1993a. "The Chemist's War. The Impact of World War I on the American Chemical Profession." *Bulletin for the History of Chemistry* 13–14:40–47.

Rhees, David J. 1993b. "Corporate Advertising, Public Relations and Popular Exhibits: The Case of Du Pont." *History and Technology* 10:67–75.

Rogers, Everett M. 2003. *Diffusion of Innovations*, 5th ed. New York: Free Press.

Rosenberg, Charles E. 1976. *No Other Gods. On Science and American Social Thought*. Baltimore/London: The Johns Hopkins University Press.

Rosenberg, Nathan, and Richard R. Nelson. 1996. "The Roles of Universities in the Advance of Industrial Technology." In *Engines of Innovation. U.S. Industrial Research at the End of an Era*, edited by Richard S. Rosenbloom and William J. Spencer, 87–109. Boston, MA: Harvard Business School Press.

Ross, Malcolm. 1931. "Invention Factory." *The New Yorker*, November 28, 1931. http://www.newyorker.com/magazine/1931/11/28/invention-factory?mbid=social_twitter.

Rothenberg, Marc, ed. 2001. *The History of Science in the United States. An Encyclopedia*. New York & London: Garland Publishing, Inc.

Røvik, Kjell Arne. 2002. "The Secrets of the Winners: Management Ideas that Flow." In *The Expansion of Management Knowledge. Carriers, Flows, and Sources*, edited by Kerstin Sahlin-Andersson and Lars Engwall, 114–144. Stanford, CA: Stanford University Press.

Russo, Arturo. 1981. "Fundamental Research at Bell Laboratories: The Discovery of Electron Diffraction." *Historical Studies in the Physical Sciences* 12 (1):117–160.

Sahlin, Kerstin, and Linda Wedlin. 2008. "Circulating Ideas: Imitation, Translation and Editing." In *The SAGE Handbook of Organizational Institutionalism*, edited by Royston Greenwood, Christine Oliver, Roy Suddaby, and Kerstin Sahlin-Andersson, 218–242. London: SAGE.

Sahlin-Andersson, Kerstin. 1996. "Imitating by Editing Success: The Construction of Organizational Fields." In *Translating Organizational Change*, edited by Barbara Czarniawska and Guje Sevón. Berlin: de Gruyter.

Saltzman, Martin D. 2006. "Chemical Societies and Their Journals: What Can Be Learned about the State of Chemistry from an Analysis of the First Volumes of These Journals." *Bulletin for the History of Chemistry* 31 (1):19–27.

Sandberg, Jörgen. 2005. "How Do We Justify Knowledge Produced Within Interpretive Approaches?" *Organizational Research Methods* 8 (1):41–68.

Schmidt, Vivien A. 2008. "Discursive Institutionalism: The Explanatory Power of Ideas and Discourse." *Annual Review of Political Science* 11:303–326.

Schwab-Trapp, Michael. 2010. "Methodische Aspekte der Diskursanalyse. Probleme der Analyse diskursiver Auseinandersetzungen am Beispiel der deutschen Diskussion über den Kosovokrieg." In *Handbuch Sozialwissenschaftliche*

Diskursanalyse. Band 2: Forschungspraxis, edited by Reiner Keller, Andreas Hirseland, Werner Schneider, and Willy Viehöver, 171–196. Wiesbaden: VS.
Scott, W. Richard. 1994. "Conceptualizing Organizational Fields." In *Systemrationalität und Partialinteresse*, edited by Hans-Ulrich Derlien, Uta Gerhardt, and Fritz W. Scharpf, 203–221. Baden-Baden: Nomos.
Scott, W. Richard. 1995. *Institutions and Organizations*. London: SAGE.
Secord, Anne. 1994. "Science in the Pub: Artisan Botanists in Early Nineteenth-Century Lancashire." *History of Science* 32 (3):269–315.
Serres, Michel. 1992. *Übersetzung*. Berlin: Merve.
Shapin, Steven. 1988. "The House of Experiment in Seventeenth-Century England." *Isis* 79 (3):373–404.
Shapin, Steven. 2008. *The Scientific Life: A Moral History of a Late Modern Vocation*. London/Chicago: University of Chicago Press.
Shils, Edward. 1979. "The Order of Learning in the United States: The Ascendancy of the University." In *The Organization of Knowledge in Modern America, 1860-1920*, edited by Alexandra Oleson and John Voss, 19–47. London/Baltimore: Johns Hopkins University Press.
Slosson, Edwin E. 1919. *Creative Chemistry*. New York: The Century Co.
Smith Jr., John Kenly. 1990. "The Scientific Tradition in American Industrial Research." *Technology and Culture* 31 (1):121–131.
Smith Jr., John Kenly, and David Hounshell. 1985. "Wallace H. Carothers and Fundamental Research at Du Pont." *Science Technology Human Values* 229 (4712):436–442.
Solli, Rolf, Peter Demediuk, and Robert Sims. 2005. "The Namesake: On Best Value and Other Reformmarks." In *Global Ideas. How Ideas, Objects and Practices Travel in the Global Economy*, edited by Barbara Czarniawska and Guje Sevón, 30–46. Copenhagen: Liber.
Soule, Sarah A. 1999. "The Diffusion of an Unsuccessful Innovation." *Annals of the American Academy of Political and Social Science* 566 (1):120–131.
Spero, Ellan F. 2014. *Institutes for Innovation: The Emergence of Academic-Industrial Cooperation and Narratives of Progress in the Early 20th Century*. Cambridge, MA: Massachusetts Institute of Technology.
Steen, Kathryn. 2014. *The American Synthetic Organic Chemicals Industry. War and Politics, 1910-1930*. Chapel Hill, NC: University of North Carolina Press.
Stiles, William B. 1993. "Quality Control in Qualitative Research." *Clinical Psychology Review* 13:593–618.
Strang, David, and John W. Meyer. 1993. "Institutional Conditions for Diffusion." *Theory and Society* 22 (4):487–511.
Strang, David, and Sarah A. Soule. 1998. "Diffusion in Organizations and Social Movements: From Hybrid Corn to Poison Pills." *Annual Review of Sociology* 24:265–290.
Sturchio, Jeffrey L. 1981. *Chemists and Industry in Modern America: Studies in the Historical Application of Science Indicators*, PhD Dissertation. Pennsylvania: University of Pennsylvania.
Sturchio, Jeffrey L. 1984. "When Industrial Research was Young: Chemistry and Corporate Strategy at Du Pont." *Research Management* 27 (1):10–18.
Sturchio, Jeffrey L. 1985. *Corporate History and the Chemical Industries*. Philadelphia, PA: Center for the History of Chemistry.

Tardy, Christine M. 2011. "Genre Analysis." In *The Bloomsbury Companion to Discourse Analysis*, edited by Ken Hyland and Brian Paltridge. London: Bloomsbury.

Taylor, Dorceta E. 2016. *The Rise of the American Conservation Movement*. Durham/London: Duke University Press.

Taylor, Kenneth L. 1976. "Two Centuries of Chemistry." In *Issues and Ideas in America*, edited by Benjamin J. Taylor and Thurman J. White. Norman, Oklahoma: University of Oklahoma Press.

Thackray, Arnold, Jeffrey L. Sturchio, P. Thomas Carroll, and Robert Bud. 1985. *Chemistry in America 1876-1976*. Dordrecht/Boston/Lancaster: D. Reidel Publishing Company.

Tobey, Ronald C. 1971. *The American Ideology of National Science, 1919-1930*. Pittsburgh, PA: University of Pittsburgh Press.

Tolbert, Pamela S., Robert J. David, and Wesley D. Sine. 2011. "Studying Choice and Change: The Intersection of Institutional Theory and Entrepreneurship Research." *Organization Science* 22 (5):1332–1344.

Tolbert, Pamela S., and Lynne G. Zucker. 1983. "Institutional Sources of Change in the Formal Structure of Organizations: The Diffusion of Civil Service Reform, 1880-1935." *Administrative Science Quarterly* 28:22–39.

Trescott, Martha Moore. 1981. *The Rise of the American Eletrochemicals Industry, 1880-1910*. Westport/London: Greenwood Press.

Union Carbide Corporation. 1976. *Our History*. New York: Public Relations Department, Union Carbide Corporation.

Usselman, Steven W. 1992. "From Novelty to Utility: George Westinghouse and the Business of Innovation during the Age of Edison." *Business History Review* 66 (2):251–304.

Van Antwerpen, F.J. 1980. "The Origins of Chemical Engineering." In *History of Chemical Engineering*, edited by William F. Furter, 1–14. Washington, DC: American Chemical Society.

Van Antwerpen, F.J., and Sylvia Fourdrinier. 1958. *High Lights: The First Fifty Years of the American Institute of Chemical Engineers*. New York: American Institute of Chemical Engineers.

van Dijk, Teun A., ed. 1985a. *Handbook of Discourse Analysis. Volume 1. Disciplines of Discourse*. London: Academic Press

Van Dijk, Teun A. 1985b. *Handbook of Discourse Analysis. Volume 2. Dimensions of Discourse*. London: Academic Press.

Van Dijk, Teun A. 1997. "The Study of Discourse." In *Discourse as Structure and Process*, edited by Teun A. Van Dijk, 1–34. London: SAGE Publications.

Vandome, Robin. 2013. "The Advancement of 'Science': James McKeen Cattell and the Networks of Prestige and Authority, 1894-1915." *American Periodicals* 23 (2):172–187.

Wæraas, Arild, and Jesper Agger Nielsen. 2016. "Translation Theory 'Translated': Three Perspectives on Translation in Organizational Research." *International Journal of Management Reviews* 18:236–270.

Walgenbach, Peter, and Renate E. Meyer. 2008. *Neoinstitutionalistische Organisationstheorie*. Stuttgart: W. Kohlhammer.

Wasserman, Neil H. 1985. "The Changing Technological Culture at AT&T." *Research Management* 28 (5):35–43.

Weart, Spencer. 1976. "The Rise of 'Prostituted' Physics." *Nature* 262:13–17.

Weart, Spencer. 1979. "The Physics Business in America, 1919-1940: A Statistical Reconnaissance." In *The Sciences in the American Context: New Perspectives*, edited by Nathan Reingold, 295–358. Washington, DC: Smithsonian Institution Press.

Weingart, Peter. 2003. *Wissenschaftssoziologie*. Bielefeld: Transcript.

Wejnert, Barbara. 2002. "Integrating Models of Diffusion of Innovations: A Conceptual Framework." *Annual Review of Sociology* 28:297–326.

West, Clarence J. 1930. *Five Years of Research in Industry. 1926-1930. A Reading List of Selected Articles from the Technical Press*. New York: National Research Council, Division of Engineering and Industrial Research.

Whalley, Peter. 1991. "The Social Practice of Independent Inventing." *Science Technology Human Values* 16 (2):208–232.

White, Paul. 2016. "The Man of Science." In *A Companion to the History of Science*, edited by Bernard Lightman, 153–163. West Sussex, UK: John Wiley & Sons, Ltd.

Whitehead, Don. 1968. *The Dow Story. The History of the Dow Chemical Company*. New York: McGraw-Hill Book Company.

Whitley, Richard. 2000. *The Intellectual and Social Organization of the Sciences*, 2nd ed. Oxford: Oxford University Press.

Whitrow, Magda. 1976. *ISIS Cumulative Bibliography: A Bibliography of the History of Science formed from ISIS Critical Bibliographies 1-90. Subjects*, Vol. 3. London: Mansell.

Wiebe, Robert H. 1967. *The Search for Order 1877-1920*. New York: Hill and Wang.

Wilkinson, Norman B. 1984. *Lammot Du Pont and the American Explosives Industry 1850-1884*. Virginia: University Press of Virginia.

Williams, S. R. 1921. "The Spirit of Research." *Science* 54 (1405):538–539.

Wise, George. 1983. "Ionists in Industry: Physical Chemistry at General Electric, 1900-1915." *Isis* 74 (1):6–21.

Wise, George. 1985. *Willis R. Whitney, General Electric, and the Origins of U.S. Industrial Research*. New York: Columbia University Press.

Wise, George. 1987. "The First Corporate R&D Organization." *Chemtech: The Innovator's Magazine* (17):394–396.

Wise, George. 1992. "Inventors and Corporations in the Electrical Industry, 1890-1940." In *Inventive Minds. Creativity in Technology*, edited by Robert J. Weber and David N. Perkins, 291–310. New York/Oxford: Oxford University Press.

Wise, George. 1997. "A New Role for Professional Scientists in Industry: Industrial Research at General Electric, 1900-1916." In *Technology & American History. A Historical Anthology from Technology & Culture*, edited by Stephen H. Cutcliffe and Terry S. Reynolds, 217–238. London/Chicago: University of Chicago Press.

Wodak, Ruth. 2001. "The Discourse-Historical Approach." In *Methods of Critical Discourse Analysis*, edited by Ruth Wodak and Michael Meyer, 63–94. London: SAGE Publications.

Wodak, Ruth, and Michael Meyer, eds. 2001. *Methods of Critical Discourse Analysis*. London: SAGE Publications.

Wood, Linda A., and Rolf O. Kroger. 2000. *Doing Discourse Analysis. Methods for Studying Action in Talk and Text*. Thousand Oaks: Sage Publications.
Woodlief, Annette M. 1981. "Science in Popular Culture." In *Handbook of American Popular Culture*, edited by Thomas Inge. Westport, CT: Greenwood Press.
Wooten, Melissa, and Andrew J. Hoffman. 2008. "Organizational Fields: Past, Present and Future." In *The SAGE Handbook of Organizational Institutionalism*, edited by Royston Greenwood, Christine Oliver, Roy Suddaby, and Kerstin Sahlin, 130–147. London: SAGE.
Zucker, Lynne G. 1977. "The Role of Institutionalization in Cultural Persistence." *American Sociological Review* 42:726–743.

Archival Sources

C.E. Kenneth Mees Correspondence. *George Eastman House Materials*. International Museum of Photography, Rochester, NY.
David A. Hounshell and John K. Smith. *Research Notes for Science and Corporate Strategy (Accession 1850)*. Hagley Museum and Library, Wilmington, DE.
George Eastman Correspondence. *Eastman Legacy Collections*. George Eastman Museum, Rochester, NY.
George Eastman Correspondence. Outgoing. Business (GEC), Personal (GECP).
George Eastman Correspondence. Incoming.
Kodak Historical Collection #0003. River Campus Libraries. University of Rochester, Rochester, NY.
Series 2. Corporate Papers.
Subseries 1. Correspondence.
Subseries 3. Executive and Management Papers.
Subseries 4. History and Development of Eastman Kodak Company.
Subseries 7. Meeting Minutes and Reports.
Series 6. Divisions.
Series 7. Kodak Research Laboratories
Records of the American Institute of Chemists. Science History Institute, Philadelphia, PA.
Records of the Chemists' Club. Science History Institute, Philadelphia, PA.

Discourse Analysis Corpus

"The Career of a Chemist – A Rejoinder." 1895. *Scientific American* 72 (14):211.
"Address of the President." 1900. *Science* 12 (288):12–15.
"Chemical Industry in Germany in 1901." 1902. *Science* 16 (397):237–238.
"Research Laboratory at Niagara Falls." 1902. *Electrochemical Industry* 1 (9):340.
"Grants Made by the Carnegie Institution." 1903. *Science* 18 (469):801–822.
"A Proposed Society of Chemical Engineers." 1907. *Electrochemical and Metallurgical Industry* 5 (7):285.
"Why Not 'The American Society of Chemical Enginers?'." 1907. *Chemical Engineer* 5 (5):227–228.
"The AIChE." 1908. *Chemical Engineer* 8 (1):1–21.

"Notes and Comments: Room for Research Work in Chemical Engineering." 1908. *Chemical Engineer* 7 (1):29–31.

"Notes and Comments: The German Chemical Industry." 1908. *Chemical Engineer* 8 (3):167–170.

"Some Thoughts on the Organization of the 'American Institute of Chemical Engineers'." 1908. *Transactions of the American Institute of Chemical Engineers* 1:35–43.

"The Status of Electrochemical Industries." 1908. *Electrochemical and Metallurgical Industry* 6 (1):2.

"Justus von Liebig and the First Laboratory." 1909. *Popular Science Monthly* 74 (34):615–619.

"The Conservation of Research." 1911. *Journal of Industrial and Engineering Chemistry* 3 (10):713–714.

"Facilities for Industrial Research." 1911. *Journal of Industrial and Engineering Chemistry* 3 (11):797–799.

"The Industrial Corporation and the Inventor." 1911. *Scientific American* 105 (21):448–449.

"Industrial Research." 1911. *Journal of Industrial and Engineering Chemistry* 3 (2):62–63.

"The New Research Laboratory Building of the American Rolling Mill Co., Middletown, Ohio." 1911. *Journal of Industrial and Engineering Chemistry* 3 (6):415–417.

"Ten Years of the Carnegie Institution." 1912. *Popular Science Monthly*:308–310.

"Engineers, Practical Men, and Theories." 1913. *Metallurgical and Chemical Engineering* 11 (4):174.

"Research Institutions." 1913. *Popular Science Monthly* 82 (4):413–415.

"The Chemist and the Banker." 1916. *Metallurgical and Chemical Engineering* 14 (4):174–175.

"The Research Corporation." 1917. *Science* 46 (1180):131–132.

"Science and Industry." 1917. *Science* 46 (1178):89.

"Scientific Research and the Electrical World." 1917. *Science* 45 (1162):331–332.

"Where Are the Leaders?" 1918. *Journal of Industrial and Engineering Chemistry* 10 (3):172.

"Shall I Educate My Boy to Be a Chemist?" 1921. *Chemical Age* 29 (5):181–182.

"American Institute of Chemistry." 1922. *Chemical Age* 30 (2):75–76.

"Cogitations on the Chemist." 1922. *Chemical Age* 30 (1):18.

Ames, J.S. 1918. "The Trained Man of Science and the War." *Science* 48 (1243):401–410.

Armstrong, E.F. 1923. "Chemistry and Industry." *Journal of the Society of Chemical Industry* 42 (25):603–612.

Auchy, George. 1909. "An American Institute of Chemistry." *Journal of Industrial and Engineering Chemistry* 1 (2):64–65.

Bacon, Raymond F. 1916. "The War and American Chemical Industry." *Chemical Engineer* 23 (6):221–229.

Bacon, Raymond F. 1917a. "The Professional Status of the Chemist." *Journal of Industrial and Engineering Chemistry* 9 (8):797–801.

Bacon, Raymond F. 1917b. "The Remuneration of Industry by Research." *Scientific American* 116 (11):281.

Bacon, Raymond F., C.E. Kenneth Mees, William H. Walker, M.C. Whitaker, and Willis R. Whitney. 1917. "Research in Industrial Laboratories." *Science* 45 (1150):34–39.

Baekeland, L.H. 1912. "Protection of Intellectual Property in Relation to Chemical Industry." *Transactions of the American Institute of Chemical Engineers* 5:19–42.

Baekeland, L.H. 1917. "The Future of Chemical Industry in the United States." *Journal of Industrial and Engineering Chemistry* 9 (11):1020–1022.

Bailey, E.H.S. 1911. "The Chemist as a Conservationist." *Science* 34 (883):748–750.

Baker, Ira O. 1900. "Engineering Education in the United States at the End of the Century." *Science* 12 (305):666–674.

Bancroft, Wilder D. 1912. "Physical-Industrial Chemistry." *Journal of Industrial and Engineering Chemistry*:79–80.

Bancroft, Wilder D. 1918. "Scientific Research After the War." *Transactions of the American Electrochemical Society* 33 (1):75–79.

Bancroft, Wilder D. 1922. "The Field for Chemists." *Journal of Industrial and Engineering Chemistry* 14 (2):156–158.

Bauer, L.A. 1909. "The Instruments and Methods of Research." *Popular Science Monthly* 74 (8):184–201.

Benner, R.C. 1912. "Research on the Smoke Problem at the Department of Industrial Research of the University of Pittsburgh." *Science* 34 (913):977–979.

Bessey, Charles E. 1896. "Science and Culture." *Science* 4 (83):121–124.

Bogert, Marston Taylor. 1913. "The Chemical Industries and the Universities." *Journal of the Society of Chemical Industry* 32 (14):720–725.

Bogert, Marston Taylor. 1917a. "The National Research Council and Its Chemistry Committee. A Preliminary Statement to the Chemists of the United States." *Journal of the American Chemical Society* 29 (5):841–855.

Bogert, Marston Taylor. 1917b. "National Research Council, Chemistry Committee Second Report " *Journal of Industrial and Engineering Chemistry* 9 (11):1009–1013.

Bolling, Randolph, and A.E. Maze. 1922. "Chemists in Public Life? Why." *Chemical and Metallurgical Engineering* 10 (9):485.

Booth, W.M.M. 1912. "The Chemical Engineer and Industrial Efficiency." *Transactions of the American Institute of Chemical Engineers* 5:184–196.

Brogdon, J.S. 1912. "The Analyst versus the Chemist." *Journal of Industrial and Engineering Chemistry* 4 (9):684–685.

Brooks, W.K. 1895. "Science or Poetry." *Science* 2 (40):437–440.

Bruce, George H. 1917. "America's Chemical Progress." *Chemical Engineer* 25 (2):126.

Burgess, C.F. 1911. "The Efficiency of College Training of Men for the Chemical Industries." *Journal of the American Chemical Society* 33 (4):611–618.

Burgess, C.F. 1923. "The Market for Chemical Ideas." *Journal of Industrial and Engineering Chemistry* 15 (10):997–1000.

Bushee, Frederick A. 1911. "Science and Social Progress." *Popular Science Monthly* 79 (15):236–251.

Campbell, Edward L. 1921. "The Consulting Chemist, Chemical Publicity and Some Problems of the Profession." *Chemical Age* 29 (1):15–16.

Cannon, W.B. 1911. "The Career of the Investigator." *Science* 34 (864):65–72.

Carhart, Henry S. 1895. "The Educational and Industrial Value of Science" *Science* 1 (15):393–402.

Carpenter, Charles C. 1917. "Chemistry and Engineering. A Timely Message to Industrial Chemists." *Chemical Engineer* 25:18–21.

Carty, John J. 1916. "The Relation of Pure Science to Industrial Research." *Science* 44 (1137):511–518.

Carty, John J. 1916. "The Relation of Pure Science to Industrial Research." *Science* 44 (1137):511–518.

Chamberlin, T.C., Franz Boas, William Morton Wheeler, Conway MacMillan, and Hugo Münsterberg. 1903. "How Can Endowments be Used Most Effectively for Scientific Research" *Science* 17 (432):571–583.

Chandler, C.F. 1900. "President's Address: Chemical and Technical Education in the United States." *Journal of the Society of Chemical Industry* 19 (7):591–594.

Chandler, C.F. 1921. "Presentation Address." *Journal of Industrial and Engineering Chemistry* 13 (2):160–161.

Choate, Parker C. 1917. "Research and Industry." *Metallurgical and Chemical Engineering* 16 (5):244–245.

Clark, Laurance T. 1917a. "A Luxury for the Industrial Chemist." *Chemical Engineer* 26 (9):323–324.

Clark, Laurance T. 1917b. "Business Chemistry." *Chemical Engineer* 25 (1):1.

Clarke, F.W. 1902. "The Development of Chemistry." *Journal of the American Chemical Society* 24 (2):117–138.

Clarke, F.W. 1916. "The Interrelations of Pure and Applied Chemistry." *Science* 43 (1104):257–263.

Clayton, H.H. 1902. "The Endowment of Research." *Science* 15 (374):351–353.

Comey, Arthur M. 1912. "Certain Phases of Technical Chemical Research." *Chemical Engineer* 15 (3):109–112.

Coolidge, Walter H. 1921. "The Relation of Chemical Training to Industry." *Science* 54 (1399):367–371.

Cottrell, F.G. 1912. "The Research Corporation, An Experiment in Public Administration of Patent Rights." *Journal of Industrial and Engineering Chemistry* 4 (12):864–867.

Coulter, John M. 1900. "The Mission of Science in Education." *Science* 12 (295):281–293.

Crowell, John Franklin. 1909. "Science and Investment." *Science* 29 (745):561–566.

de Long, C.R. 1922. "Service of Department of Commerce to Chemical Industry." *Chemical Age* 30 (10):431–433.

Derick, C.G. 1922. "The Relation of Research to Production Efficiency." *Chemical and Metallurgical Engineering* 27 (9):425–427.

Doolittle, C.L. 1902. "Scientific Research." *Science* 15 (387):841–848.

Duisberg, Carl. 1896. "The Education of Chemists." *Journal of the Society of Chemical Industry* 15 (6):427–432.

Duisberg, Carl. 1912. "The Latest Achievements and Problems of the Chemical Industry." *Chemical Engineer* 16 (6):221–230.

Duncan, Robert Kennedy. 1907a. *The New Knowledge. A Popular Account of the New Physics and the New Chemistry in their Relation to the New Theory of matter*. London: Hodder and Stoughton.

Duncan, Robert Kennedy. 1907b. *Chemistry for Commerce*. London/New York: Harper & Brothers.

Duncan, Robert Kennedy. 1912. "Progress in Industrial Fellowships." *Science* 35 (888):19–21.

du Puy, William A. 1912. "A Professional Inventor." *Scientific American* 106 (13):292–295.

Eisenschiml, Otto. 1922a. "Some Economic Aspects of the Practice of Chemistry." *Chemical Age* 30 (3):139–140.

Eisenschiml, Otto. 1922b. "Mid-West Opinion on the Formation of An Economic Organization of Chemists." *Chemical Age* 30 (5):233–235.

Eisenschiml, Otto. 1922c. "Is a Commercial War between Chemists and Business Men Coming?" *Journal of Industrial and Engineering Chemistry* 14 (5):356.

Emich, Friedrich. 1900. "The Accuracy of the Experimental Methods of the Chemist." *Science* 11 (273):443–449.

Ferguson, William C. 1912. "A Plan for Organized Research and Analytical Chemistry in Successful Chemical Manufacturing." *Journal of Industrial and Engineering Chemistry* 4 (12):905–908.

Flinn, Alfred D. 1921. "The Technical School and Industrial Research." *Chemical Age* 29 (12):487.

Fullerton, George Stuart. 1900. "The Faith of Science." *Science* 12 (303):586–589.

Garvan, Francis P. 1921. "Chemistry and the State." *Journal of Industrial and Engineering Chemistry* 13 (10):866–874.

Gilfin, J. Elliott. 1900. "Notes on Industrial Chemistry." *American Chemical Journal* 24 (6):530–534.

Gillett, H.W. 1913. "The Chemist and Scientific Management." *Journal of Industrial and Engineering Chemistry* 5 (7):593–601.

Goode, G. Brown. 1895. "America's Relation to the Advance of Science." *Science* 1 (1):4–9.

Grosh, Daniel M. 1913. "What's the Matter with the American Chemist?" *Journal of Industrial and Engineering Chemistry* 5 (8):692–693.

Hadfield, Robert. 1916. "The Training of Our Captains of Industry." *Metallurgical and Chemical Engineering* 15 (4):171–172.

Hale, George Ellery. 1918. "Industrial Research and National Welfare." *Science* 48 (1247):505–507.

Hale, William J. 1921. "The Immediate Needs of Chemistry in America." *Journal of Industrial and Engineering Chemistry* 13 (5):460–465.

Halstead, George Bruce. 1895a. "Original Research and Creative Authorship the Essence of University Teaching." *Science* 1 (8):203–207.

Halstead, George Bruce. 1895b. "Pasteur as Illustration of Modern Science." *Science* 2 (45):611–613.

Hemingway, Frank. 1917. "Organization of Chemical Industries." *Transactions of the American Institute of Chemical Engineers* 10:356–366.

Herty, Charles H. 1917. "The Battle of the American Chemical Industries." *Journal of Industrial and Engineering Chemistry* 9 (1):1005.

Hill, Charles W. 1918. "Efficiency in Industrial Research." *Chemical and Metallurgical Engineering* 18 (4):182–184.

Hillebrand, W.F. 1907. "The Present and Future of the American Chemical Society." *Journal of the American Chemical Society* 29 (2):1–18.

Howe, H.E. 1902. "Metallurgical Laboratories." *Science* 15 (385):761–766.

Howe, H.E. 1919. "The Organization of Scientific and Industrial Research at Home and Abroad." *Bulletin of the National Research Council* 1 (1):21–43.

Hyde, Edward P. 1909. "The Physical Laboratory of the National Electric Lamp Association." *Science* 30 (782):907–909.

Hyde, Edward P. 1922. "Research from the Business Man's Standpoint." *Chemical Age* 30 (7):289–291.

Jacobson, C. Alfred. 1916. "The Importance of Scientific Research to the Industries." *Science* 44 (1135):456–459.

Kellogg, Vernon. 1921. "The University and Research." *Chemical Age* 29 (7):274–276.

Layton, Caleb R. 1921. "The Chemical Age." *Chemical Age* 29 (7):260.

Little, A.D. 1907. "The Chemist and the Community." *Science* 25 (643):647–653.

Little, A.D. 1908. "The Dependence of Civilization on Chemistry." *Electrochemical and Metallurgical Industry* 6 (3):96.

Little, A.D. 1909. "The Untilled Field of Chemistry." *Journal of Industrial and Engineering Chemistry* 1 (2):105–107.

Little, A.D. 1911. "The Earning Power of Chemistry." *Chemical Engineer* 14 (3):366–372.

Little, A.D. 1913. "Industrial Research in America." *Science* 38 (984):643–656.

Little, A.D. 1916. "Industrial Research and Its Fruitage." *Chemical Engineer* 24 (2):78–79.

Little, A.D. 1923. "The Place of Chemistry in Business." *Journal of Industrial and Engineering Chemistry* 13 (5):386–391.

Little, A.D., and G.A. O'Reilly. 1917. "Chemistry and the Banker." *Metallurgical and Chemical Engineering* 17 (8):476–481.

Louis, Henry. 1917. "President's Address." *Journal of the Society of Chemical Industry* 36 (14):742–743.

Lovelace, B.F. 1921. "Relation of University Chemical Departments to Chemical Industry." *Chemical Age* 29 (9):357–359.

MacDowell, Charles H. 1922. "The Importance of Scientific Control in Increased Production Efficiency." *Chemical and Metallurgical Engineering* 27 (9):430–431.

Maclaurin, Richard C. 1916. "Science Linked With Industry." *Metallurgical and Chemical Engineering* 15 (1):43–44.

Maclaurin, Richard C., Henry P. Talbot, William H. Walker, and A.D. Little. 1916. "The Universities and the Industries." *Journal of Industrial and Engineering Chemistry* 8 (1):59–66.

MacMillan, Conway. 1895. "The Scientific Method and Modern Intellectual Life." *Science* 1 (20):537–542.

Mason, A.H. 1895. "Chairman's Address." *Journal of the Society of Chemical Industry* 14 (10):849–851.

Matthews, J. Merritt. 1902. "The Industrial Development of Indigo." *Journal of the Franklin Institute*: 423–429.

Matthews, J. Merritt. 1916. "The New Aspects of Chemical Science." *Journal of Industrial and Engineering Chemistry* 8 (12):1147–1148.

McKeen Cattell, J. 1902. "The Carnegie Institution." *Science* 16 (403):460–469.

McMurrich, J. Playfair. 1907. "Present-Day Conditions and the Responsibilities of the University." *Science* 25 (643):641–647.

McMurtrie, William. 1895. "The Relations of the Industries to the Advancement of Chemical Science." *Science* 2 (36):283–296.

McMurtrie, William. 1899. "The Condition, Prospects and Future Educational Demands of the Chemical Industries." *Science* 13 (325):441–453.

McMurtrie, William. 1901. "The Condition, Prospects and Future Educational Demands of the Chemical Industries." *Science* 13 (325):441–453.

McMurtrie, William. 1907. "Relation of Chemistry to the Industries." *Science* 26 (673):694–699.

McMurtrie, William. 1909. "Post-Graduate Study in Applied Chemistry." *Science* 29 (746):601–605.

Mees, C.E. Kenneth. 1916. "The Organization of Industrial Scientific Research." Proceedings of the American Chemical Society, 64ff.

Mees, C.E. Kenneth. 1917a. "The Production of Scientific Knowledge." *Journal of Industrial and Engineering Chemistry* 9 (12):1137–1141.

Mees, C.E. Kenneth. 1917b. "The Publication of Scientific Research." *Science* 46 (1184):237–238.

Messel, Rudolph. 1912. "Progress in Industrial Chemistry." *Chemical Engineer* 16 (5):171–174.

Mitchell, John Purroy. 1912. "Proceedings of the Thirty-First Annual Meeting." *Journal of the Society of Chemical Industry* 31 (18):853.

Montgomery, T.H. 1908. "The Practical Value of Pure Science." *Popular Science Monthly* 73:234–247.

Newcomb, Simon. 1895. "To Our Readers." *Science* 1 (1):1–2.

Nichols, William H. 1918. "Preisdent's Address. A Retrospect and an Application." *Journal of Industrial and Engineering Chemistry* 10 (10):768–772.

Norton, Thomas H. 1917. "The Organization of National Industrial Research." *Chemical Engineer* 24:201–204.

Noyes, W.A. 1917. "The Relation of Chemical Laboratories to the National Welfare." *Science* 46 (1175):1–8.

Noyes, W.A. 1922. "The Foundation for Chemical Development." *Journal of Industrial and Engineering Chemistry* 14 (9):779–780.

O'Brien, George L. 1922. "Some Conditions of the Chemist's Work." *Chemical Age* 30 (8):341–342.

Ogden, H.N. 1918. "The Purpose of Research." *Science* 48 (1248):525–532.

Palmer, Charles S. 1908. "The Training of Industrial Chemists." *Science* 28 (725):725–727.

Parker, T.J. 1909. "Editorials. The Industrial Chemist and His Journal." *Journal of Industrial and Engineering Chemistry* 1 (1):1–2.

Parmelee, E.C. 1921a. "A Question that Will Not Down." *Chemical and Metallurgical Engineering* 24 (5):193.

Parmelee, E.C. 1921b. "Research, a Capital Investment or an Operating Expense?" *Chemical and Metallurgical Engineering* 24 (5):195.

Parmelee, E.C. 1921c. "Dividing the Sheep from the Goats in Research." *Chemical and Metallurgical Engineering* 24 (4):145–146.

Parmelee, E.C. 1922. "Research is the Plow of Industry." *Chemical and Metallurgical Engineering* 27 (24):1153–1154.

Parr, S.W. 1918. "Some Developments in the Chemical Industries as a Result of War Conditions." *Science* 47 (1217):399–410.

Platt, Charles. 1894. "Current Notes on Chemistry – V. - Education of Industrial Chemists." *Science* 23 (572):37–38.
Pupin, M.I. 1895. "Tendencies of Modern Electrical Research." *Science* 2 (52):861–880.
Raiford, L. Chas. 1917. "Some Relationships of Chemistry and Life." *Science* 45 (1169):489–494.
Redman, S.L. 1922. "Chemistry in Public Life." *Chemical and Metallurgical Engineering* 27 (7):292.
Reese, Charles Lee. 1918. "Developments in Industrial Research." *Chemical and Metallurgical Engineering* 19 (4):197–199.
Reid, E. Emmett. 1922. "Ira Remsen." *Journal of Industrial and Engineering Chemistry* 14 (11):1078.
Remsen, Ira. 1894. "Prof. Ira Remsen on Chemical Laboratories." *Nature* 49 (1275):531–535.
Richards, Joseph W. 1902. "The American Electrochemical Society." *Transactions of the American Electrochemical Society* 1 (1):1–2.
Richards, Joseph W. 1903. "Conditions of Progress in Electrochemistry." *Transactions of the American Electrochemical Society* 3:59–73.
Richardson, Clifford. 1908. "A Plea for Broader Education of the Chemical Engineer." *Science* 27 (681):81–89.
Richardson, W.D. 1908a. "The Analyst, the Chemist and the Chemical Engineer." *Science* 28 (717):396–402.
Richardson, W.D. 1908b. "The Usefulness of Chemistry in the Industries." *Science* 27 (699):801–812.
Richardson, W.D. 1911. "The Lost Arts of Chemistry." *Science* 33 (849):513–520.
Robertson, T. Brailsford. 1917. "The Utilization of Patents for the Promotion of Research." *Science* 46 (1190):371–379.
Roeber, E.F. 1908. "The Chemical Engineer." *Electrochemical and Metallurgical Industry* 6 (8):309.
Roeber, E.F. 1911a. "The Position of Electrochemistry in the Chemical and Metallurgical Industries." *Metallurgical and Chemical Engineering* 9 (1):1–2.
Roeber, E.F. 1911b. "Organized Research in America." *Metallurgical and Chemical Engineering* 11 (12):665–666.
Roeber, E.F. 1912. "Nineteen Hundred and Eleven." *Metallurgical and Chemical Engineering* 10 (1):1–2.
Roeber, E.F. 1913. "Organized Research in America." *Metallurgical and Chemical Engineering* 11 (12):665–666.
Roeber, E.F. 1916. "One Function of Scientific and Engineering Societies in Democracy." *Metallurgical and Chemical Engineering* 14 (8):405–407.
Roeber, E.F. 1917. "The Chemist and the Banker." *Metallurgical and Chemical Engineering* 17 (8):449–450.
Root, Elihu. 1918. "Industrial Research and National Welfare." *Science* 48 (1248):532–534.
Russell, I.K. 1923. "The Earning Power of Research." *Journal of Industrial and Engineering Chemistry* 15 (6):645–647.
Sadtler, Samuel P. 1908. "Some Thoughts on the Organization of the "American Institute of Chemical Engineers"." *Transactions of the American Institute of Chemical Engineers* 1:35–43.

Sadtler, Samuel P. 1909. "Conservation and the Chemical Engineer." *Transactions of the American Institute of Chemical Engineers* 2:105–114.

Scholes, S.R. 1918. "Permanence as an Ideal of Research." *Journal of Industrial and Engineering Chemistry* 10 (5):390–391.

Scripture, E.W. 1895. "The Nature of Science and Its Relation to Philosophy." *Science* 1 (13):350–352.

Sedgwick, W.T. 1913. "The Interpretation of Nature and the Teaching Laboratory." *Science* 38 (971):169–174.

Seidell, L.R. 1922. "The Practice of Consulting Chemistry as a Profession in New York." *Chemical Age* 30 (9):385–386.

Sharp, Clayton H. 1917. "Independent Laboratories in the Engineering Industries." *Metallurgical and Chemical Engineering* 17 (4):167–169.

Skinner, C.E. 1922. "Selling Research." *Chemical Age* 30 (4):168.

Skinner, Harvey J. 1911. "The Debt of the Manufacturer to the Chemist" *Chemical Engineer* 14 (2):307–308.

Slosson, Edwin E. 1922. "The Human Side of Chemistry." *Journal of Industrial and Engineering Chemistry* 14 (10):887–893.

Stewart, John. 1912. "A Plan for the Support of Chemical Research and for the Better Teaching of Industrial Chemistry." *Journal of Industrial and Engineering Chemistry* 4 (9):616–619.

Stieglitz, Julius. 1917a. "The Outlook of Chemistry in the United States." *Journal of the American Chemical Society* 39 (10):2095–2110.

Stieglitz, Julius. 1917b. "American Chemical Industry and the American Chemical Society." *Journal of Industrial and Engineering Chemistry* 9 (11):1005–1006.

Stieglitz, Julius, and Charles L. Parsons. 1917. "War Service for Chemists." *Journal of Industrial and Engineering Chemistry* 9 (8):730.

Swan, Joseph Wilson. 1901. "Presidential Address." *Journal of the Society of Chemical Industry* 20 (7):660–673.

Talbot, Henry P. 1912. "The Privileges and Responsibilities of the Chemical Analyst." *Science* 35 (898):397–404.

Teeple, J.E. 1908. "Some Problems of the Industrial Chemist." *Science* 28 (715):321–328.

Teeple, J.E. 1922. "Research: Its Position in the Making of an Industry." *Journal of Industrial and Engineering Chemistry* 14 (10):904–905.

Thompson, G.W. 1913. "Chemistry as Affecting the Profitableness of Industry." *Science* 38 (988):800–806.

Thompson, G.W. 1917a. "The Human Side of the Development of Chemical Industry." *Metallurgical and Chemical Engineering* 16 (2):79–81.

Thompson, G.W. 1917b. "The Progress of American Chemical Engineering." *Journal of Industrial and Engineering Chemistry* 9 (11):1007.

Thompson, G.W. 1918. "The Importance of Chemistry in Industry." *Chemical and Metallurgical Engineering* 19 (6a):355–357.

Thurston, R.H. 1900. "Patents and the Industries." *Science* 11 (276):593–594.

Thurston, R.H. 1901. "Science and Industrial Competition." *Science* 13 (329):631–632.

Thurston, R.H. 1902. "Scientific Research: The Art of Revelation and Prophecy (Part I/II)." *Science* 16 (402/403):401–424, 445–457.

Tingle, J. Bishop. 1908. "The Training of Industrial Chemists." *Science* 28 (728):840.

Toch, Maximilian. 1909. "The Influence of Chemistry on Civilization." *Science* 30 (777):697–699.

Tucker, Samuel A. 1908. "Electrochemistry as Applied to Industrial Chemistry." *Journal of the Society of Chemical Industry* 27 (6):267–268.

Turner, William. 1900. "Address of the President of the British Association for the Advancement of Science." *Science* 12 (297):357–368.

Vaughan, Victor C. 1912. "The Philosophy of a Scientist." *Science* 36 (921):225–233.

Walker, George W. 1922. "The Training of the Industrial Chemist." *Journal of Industrial and Engineering Chemistry* 14 (8):737.

Walker, William H. 1911. "Chemical Research and Industrial Progress." *Transactions of the American Electrochemical Society* 19:17–30.

Wardenburg, F.A. 1922. "Industrial Research Must Justify Itself in Economies. Research Chemists Cannot Be Employed If Results Do Not Show a Profit." *Chemical Age* 30 (2):57–58.

Washburn, Frank S. 1916. "The Day of the Chemist." *Journal of Industrial and Engineering Chemistry* 8 (12):1148–1151.

Watkins, W.H. 1913. "What's the Matter with the American Chemist?" *Journal of Industrial and Engineering Chemistry* 5 (8):785.

Weens, J.B. 1917. "The Relation of Research to Everyday Life." *Metallurgical and Chemical Engineering* 16 (9):478–480.

Weidlein, Edward R. 1922. "The Place of Scientific Research in the Programs of Trade Associations." *Chemical Age* 30 (6):241–243.

Wesson, David. 1912. "The Chemist and the Cotton Seed Oil Industry in America." *Journal of Industrial and Engineering Chemistry* 4 (1):64–65.

Whitaker, M.C. 1911. "The Training of Chemical Engineers." *Chemical and Metallurgical Engineering* 9 (1):9–11.

Whitaker, M.C. 1912a. "The New Chemical Engineering Course and Laboratories at Columbia University." *Transactions of the American Institute of Chemical Engineers* 5:150–169.

Whitaker, M.C. 1912b. "Shall My Boy Become a Chemical Engineer?" *Scientific American* 107 (1):16.

Whitney, Willis R. 1909. "Editorials. Education." *Journal of Industrial and Engineering Chemistry* 1 (3):147–148.

Whitney, Willis R. 1911. "Research as a Financial Asset." *Journal of Industrial and Engineering Chemistry* 3 (6):429–433.

Whitney, Willis R. 1916. "Research as a National Duty." *Science* 43 (1114):629–637.

Whitney, Willis R. 1921. "The Biggest Things in Chemistry." *Journal of Industrial and Engineering Chemistry* 13 (2):161–166.

Wilbur, Ray Lyman. 1916. "Research." *Science* 44 (1123):1–4.

Wiley, H.W. 1902. "The Role of Chemistry in University Education." *Science* 16 (413):841–850.

Wilson, John Arthur. 1922. "Mechanical Improvements vs. Fundamental Research in the Leather Industry." *Chemical and Metallurgical Engineering* 27 (9):465.

Withrow, James R. 1911. "Points of View in the Teaching of Industrial Chemistry." *Journal of the American Chemical Society* 33 (4):624–628.

Withrow, James R. 1916. "The American Chemist and the War's Problems." Science 43 (1120):835–842 .

Withrow, James R. 1917. "The Relation of War to Chemistry in America." *Science* 45 (1172):595–608.

Index